AF778781

DESIGN:
ISAMU NOGUCHI AND ISAMU KENMOCHI

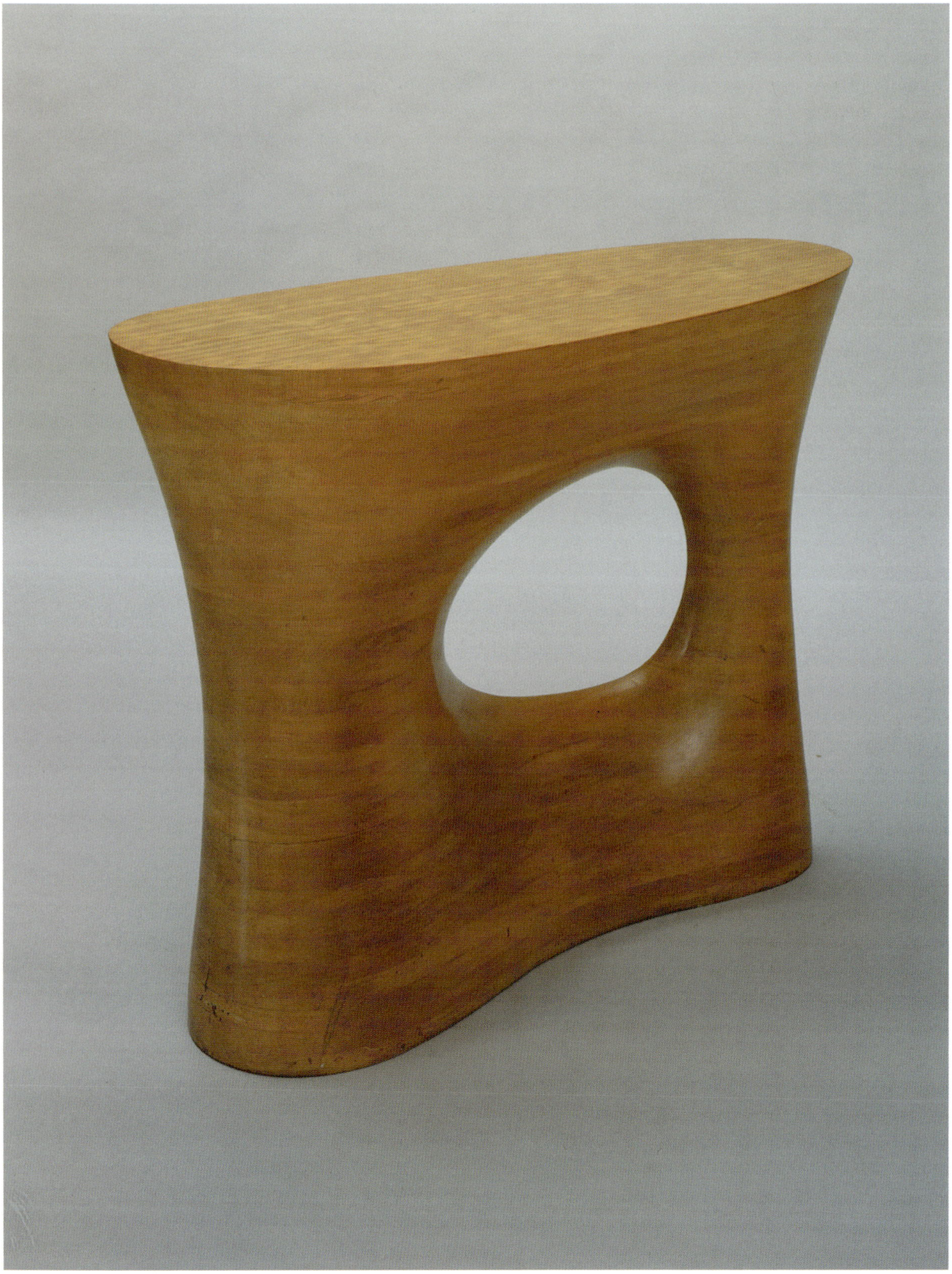

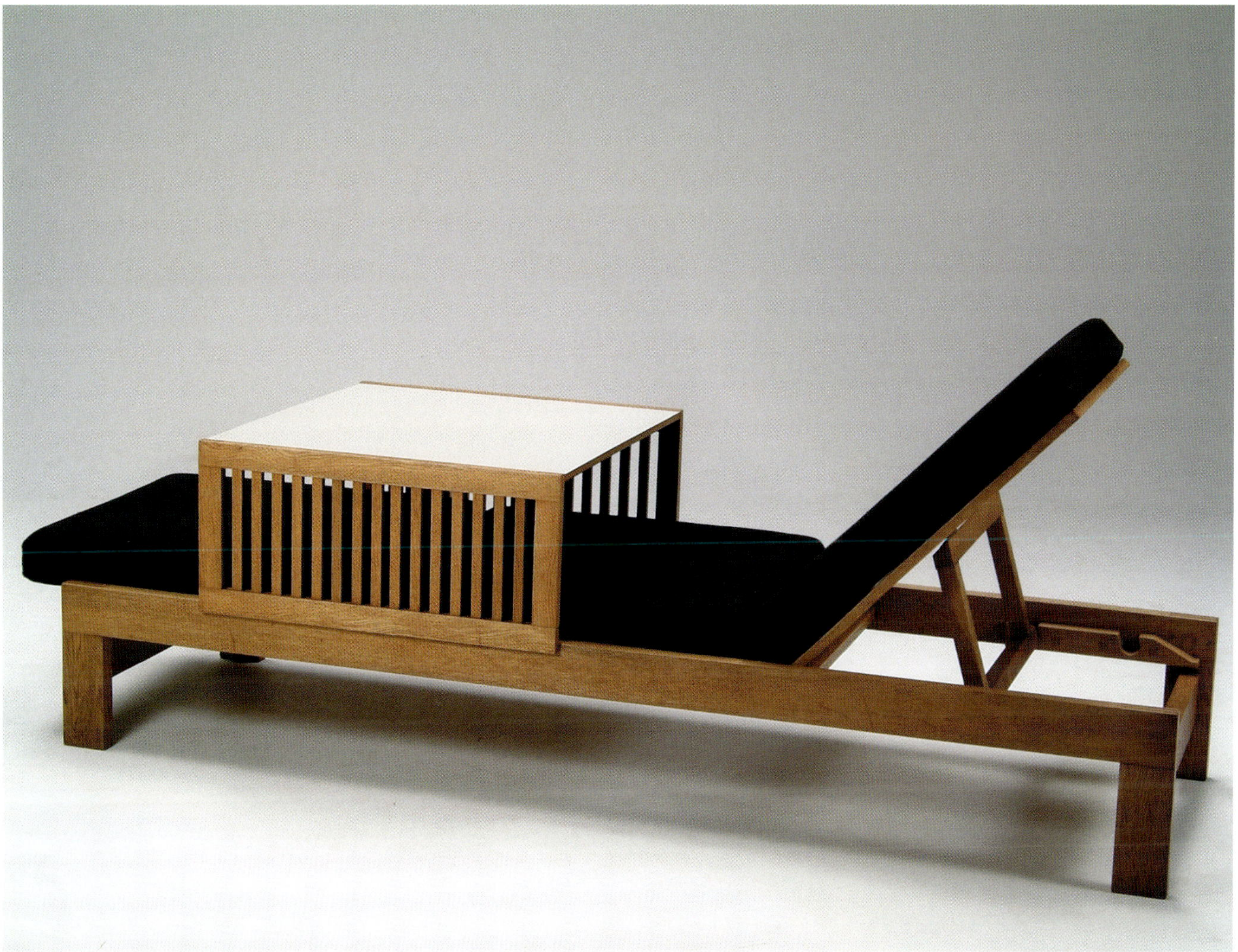

Above: Kenmochi Design Associates, Daybed, oak, 1966, 24 15/16 x 70 7/8 x 23 5/8 in. Manufactured by Yamaguchi Mokuzai Kogei. Opposite: Isamu Noguchi, Table, laminated primavera wood, ca. 1941, 30 x 41 x 16 in. Following a laminated table Noguchi made in 1939 for A. Conger Goodyear, then president of The Museum of Modern Art, New York, Noguchi made a more sculptural table in laminated wood for the architect of the Museum, Philip L. Goodwin. This unique table illustrates Noguchi's formal vocabulary at the time in relation to biomorphic forms and his own interpretation of Surrealism.

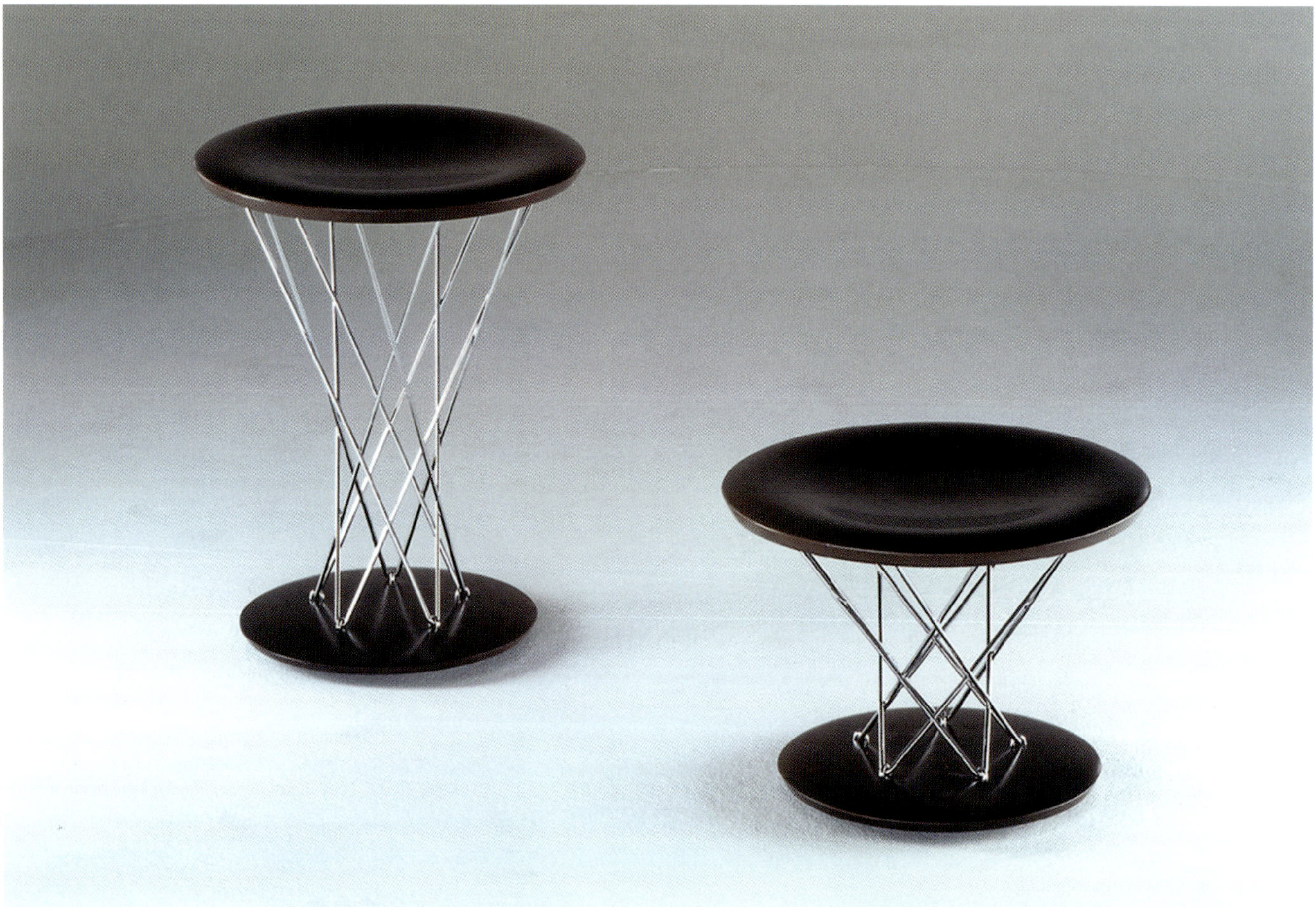

Isamu Noguchi, Rocking Stools, chrome-plated steel wire, birch, 1954, small: 8 1/2 x 13 in., large: 15 1/4 x 12 7/8 in.
Manufactured by Knoll Associates.

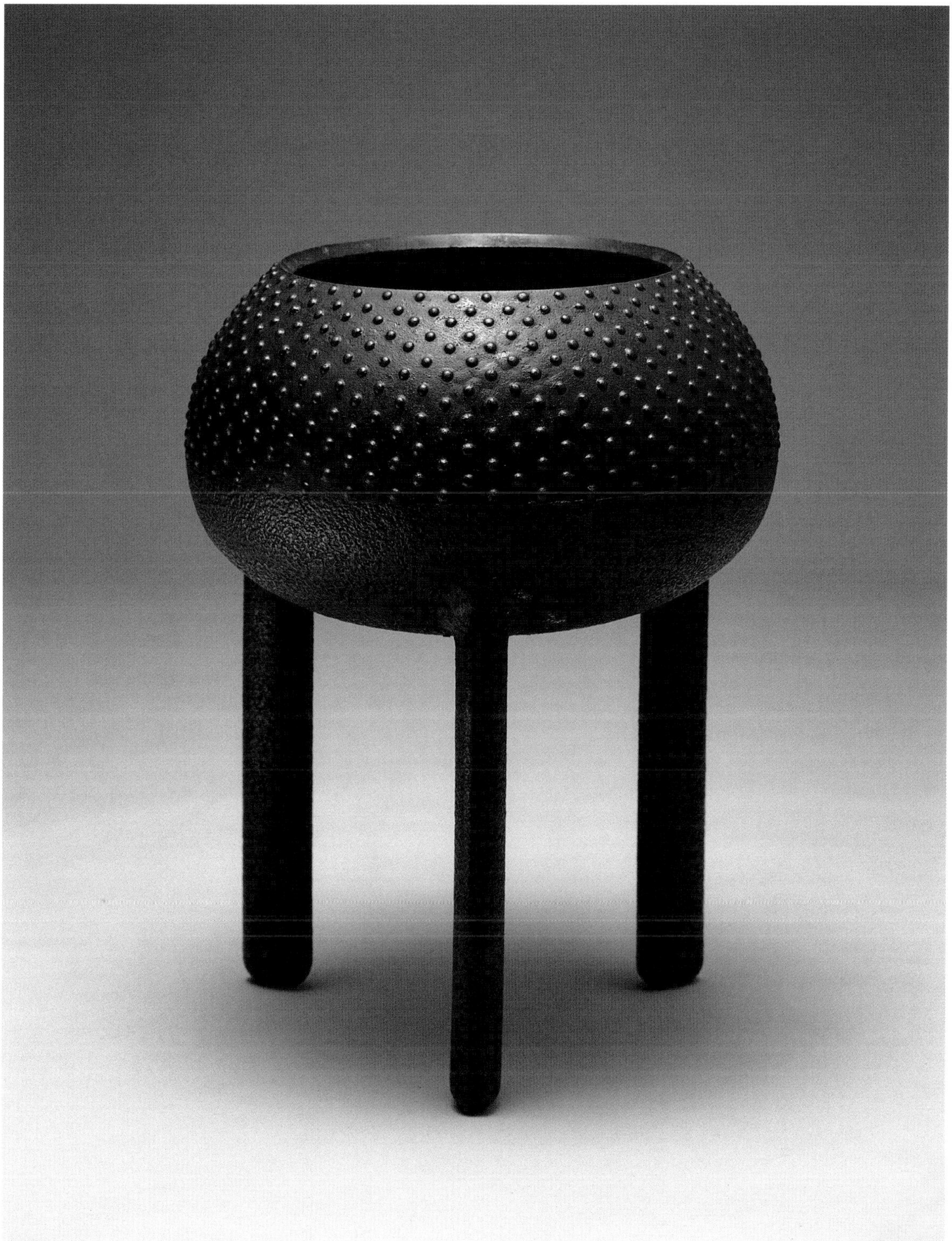

Kenmochi Design Associates, Big Ashtray, iron and lacquer, 1966, height: 19 13/16 in., diameter: 15 3/4 in. This ashtray was designed for the Kyoto International Conference Center to compliment other interior furniture items by Kenmochi Design Associates. It was cast at a foundry in Takaoka City, Yamagata Prefecture, known for its fine casting technique and a long history of maintaining and improving traditional technology.

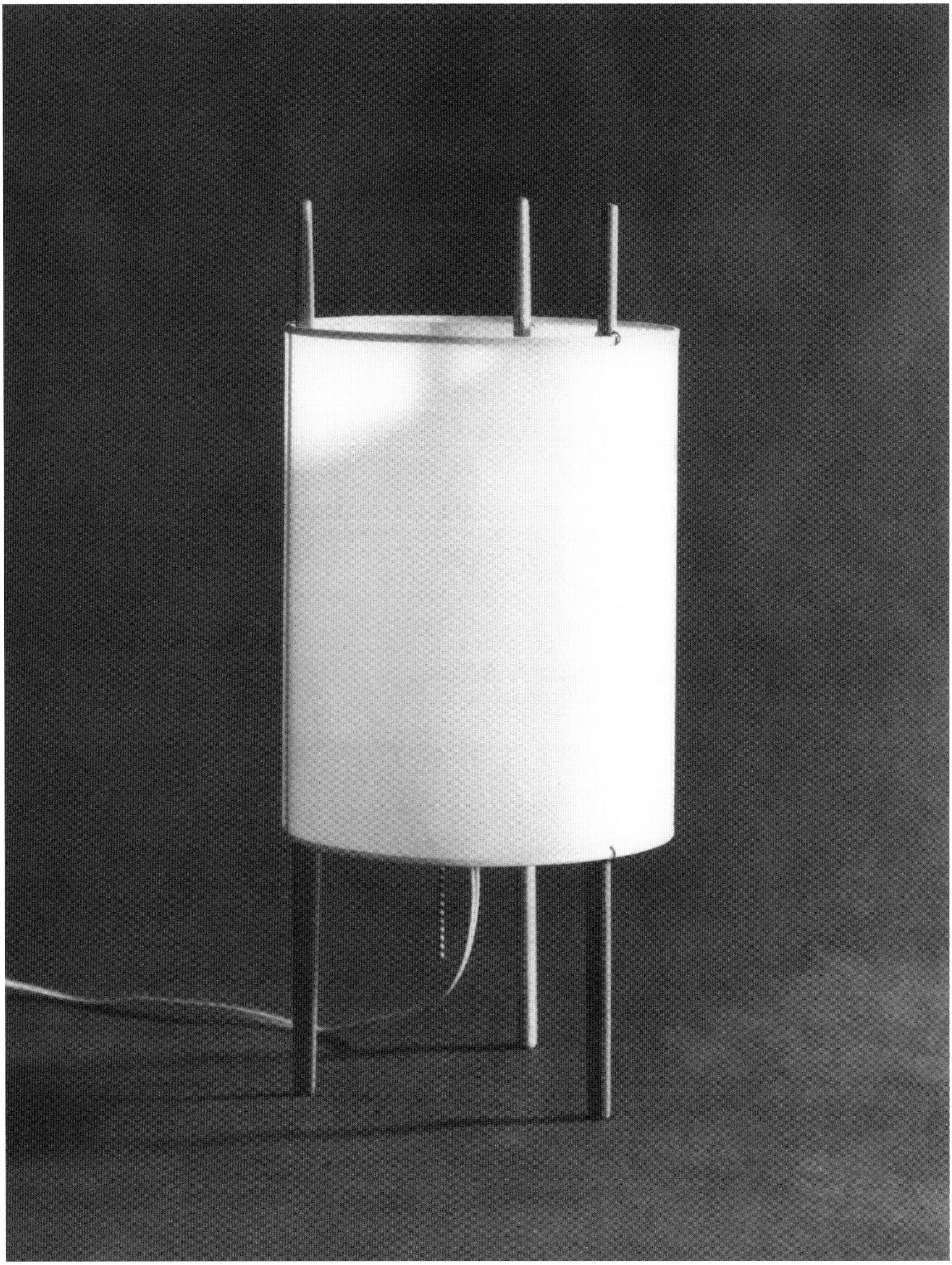

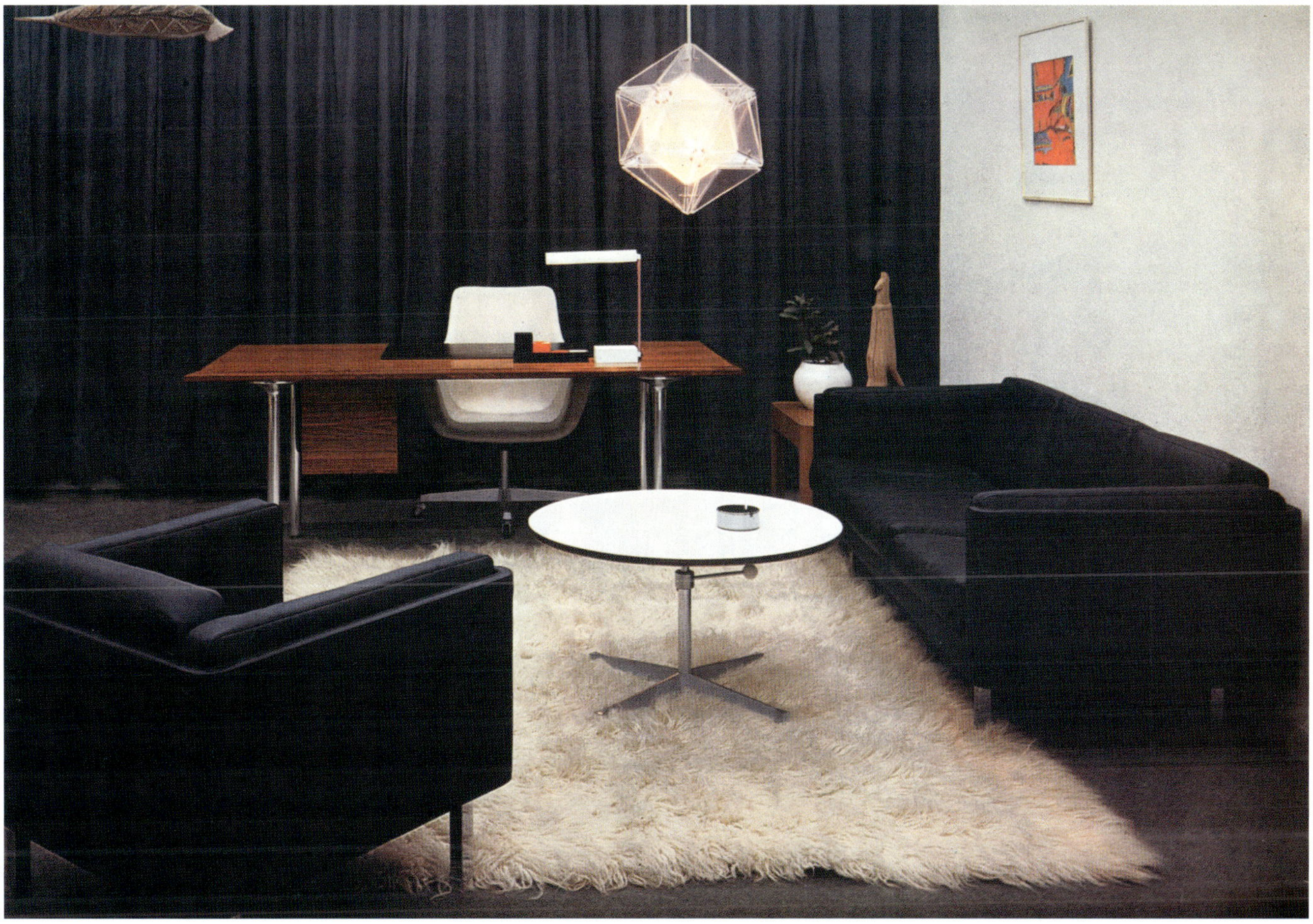

Above: Isamu Kenmochi, Showroom at Kenmochi Design Associates, 1967. This showroom of an office was designed by Kenmochi and located on the first floor of his Tokyo firm. Opposite: Isamu Noguchi, Three-Legged Cylinder Lamp, plastic and wood, ca. 1945, 15 3/4 x 7 x 7 in. Manufactured by Knoll Associates.

DESIGN:
ISAMU NOGUCHI AND ISAMU KENMOCHI

FOREWORD BY JENNY DIXON

PREFACE BY KATARINA POSCH

WITH ESSAYS BY BONNIE RYCHLAK, HITOSHI MORI,

NINA MURAYAMA, AND TETSUO MATSUMOTO

Five Ties Publishing

in association with

The Isamu Noguchi Foundation and Garden Museum

CONTENTS

FOREWORD

On first seeing a tangible example of the Bamboo Basket Chair, the result of a collaboration between the sculptor Isamu Noguchi and the industrial designer Isamu Kenmochi in 1950, one cannot help but be awed by its innate sense of engineering combined with a highly developed aesthetic sensibility. Actually, what one sees, in 2007, since the original chair no longer exists, is a carefully created facsimile, whose production process highlights the remarkable nature of the Noguchi-Kenmochi prototype made of iron and woven bamboo. The facsimile is the work of Kenmochi Design Associates in Tokyo, headed by Tetsuo Matsumoto, who worked at the prestigious, but now virtually extinct, Industrial Arts Research Institute (IARI, renamed the Industrial Arts Institute in 1952) in the mid-1950s, where Isamu Kenmochi served as "technical officer." Upon leaving the Institute, Matsumoto was invited to work with Kenmochi, who had been his mentor.

A recent visit to Kenmochi Design Associates, which features a wall lined with books on art, crafts, furniture design, and architecture in numerous languages, confirmed that Matsumoto, now in his mid-seventies and still energetic, has devoted his life to Kenmochi's designs, which continue to inspire him. Work to reproduce a chair that Noguchi and Kenmochi conceived and developed during a two-week period in August 1950 has taken Matsumoto, his colleagues—who also worked at the IAI and who remember the chair—and master craftsman and bamboo weaver Shosaku Kondo more than seven months to resolve. Clearly, these designers and artisans were also awestruck, and intrigued, by the brief initial coming together of minds and shared sensibilities that could produce a work so deceptively simple and yet so complex.

Following World War II, there was renewed interest within the Japanese arts and design community in modern ideas of form and contemporary production methods emanating from the West. The Japanese-American Isamu Noguchi was embraced in this climate, but Noguchi was also interested in integrating traditional Japanese forms and materials into his work. While in Tokyo in June 1950, he met the like-minded Japanese designer Isamu Kenmochi, who was roughly his contemporary, and a series of verbal and visual conversations began, resulting in a cross-fertilization of ideas. It was the vision of Noguchi Museum curator Bonnie Rychlak to examine, for the first time, the intersecting achievements of these two artists and friends, from their initial collaboration on the 1950 Bamboo Basket Chair, to their larger, individual production.

In Japan, Kenmochi is well known and highly regarded for his contribution to twentieth-century furniture design. In the United States, while his Rattan Round Chair and its unauthorized reproductions are familiar, the designer's name is not. In presenting the exhibition *Design: Isamu Noguchi and Isamu Kenmochi*, The Noguchi Museum is delighted to bring the work of Kenmochi to the fore and to explore the ways in which he and Noguchi inspired one another. Their shared aesthetic, in which East and West commingle, is materialized in some of the most remarkable examples of mid-twentieth-century functional design.

The Noguchi Museum is profoundly grateful to all who have contributed so generously to the creation of this exhibition and its accompanying catalogue. The exhibition marks a first for The Noguchi Museum, in that an impressive number of Japanese enterprises have offered and provided generous support and encouragement for this effort. The Honorable Motoatsu Sakurai, Ambassador of the Consulate-General of Japan in New York, and his staff were extremely helpful in the early planning stages, and assisted with important introductions. Similarly, the Honorable Robin Chandler Duke, who among her many accomplishments is a founder and Trustee of the United States–Japan Foundation, was an early proponent of the exhibition. We are especially appreciative of their early enthusiasms, which have led to further support. Through the kindness of Taro Matsui and Noboru Takemoto, All Nippon Airways Co., Ltd., provided transportation assistance for the artworks and key lenders. The United States–Japan Foundation, under the leadership of its president, George Packard, has underwritten the educational component of the exhibition. David Janes, Assistant to the President and Program Officer, also offered bountiful help. At The Japan Foundation, New York, Director General Isao Tsujimoto and his predecessor, Masaru Susaki, embraced this project with much enthusiasm from the outset and subsequently provided valuable support.

In addition, we wish to acknowledge the National Endowment for the Arts, the New York State Council on the Arts, and the New York City Department of Cultural Affairs. Without the support of these United States arts agencies, it would have been exceedingly difficult to realize the exhibition *Design: Isamu Noguchi and Isamu Kenmochi.*

Beyond indispensable financial support, there has been an outpouring of incredible kindness and cooperation extended to The Noguchi Museum in our goal to bring this exhibition to fruition. We particularly wish to express deep gratitude to our following Japanese friends and colleagues: Hitoshi Mori of the Matsudo City Board of Education, who so willingly shared with us his extensive scholarship on Kenmochi, as well as facilitated contact with key individuals and lenders; Tetsuo Matsumoto, who beyond working with immense dedication on the exhibition replicas of the prototype Bamboo Basket Chair and Bamboo Chair, loaned numerous works to the exhibition; Takuya Kida of the Crafts Gallery, The National Museum of Modern Art, Tokyo, who assisted with the history of the Akari Light Sculptures and the chronology of their production; and Suehiko Kikuchi of the Japan Industrial Arts Foundation, who provided us with many photographs from the Industrial Arts Research Institute where Noguchi and Kenmochi worked together in 1950.

The Kotobuki Corporation, Y▪M▪K Co., Ltd., and Yohko Watanabe of the Research Center for the Arts and Arts Administration, Keio University, have all been instrumental in making important examples of Kenmochi's and Noguchi's furniture available to the exhibition, providing, respectively, a replica of a Kenmochi bench from the Osaka World's Fair Expo '70, assistance with the distribution of Kenmochi's rattan furniture in North America, and supervision of the conservation of the furniture Noguchi created for *Shin Banraisha* at Keio University in Tokyo.

We also wish to thank, in Japan, all the generous lenders of works to the exhibition, those who have granted us permission to reproduce historic photographs, and those who have provided research assistance: Akita Mokko Co., Ltd.; Crafts Gallery, The National Museum of Modern Art, Tokyo; Eiji Fujimoto, Sanki Collection; Yasuhiro Ishimoto; Japan Industrial Arts Foundation; Kagawa Prefecture Gymnasium; Shufuku Kenkyujo; Kenmochi Design Associates; Suehiko Kikuchi; Kotobuki Corporation; Kyoto International Conference Center; Matsudo City Board of Education; Tetsuo Matsumoto; Musée Kiyoharu Shirabaka; The Isamu Noguchi Garden Museum, Japan; Nomadic

Studio; Ohtone Country Club; Ozeki & Co., Ltd.; Murai Osamu; Polytechnic University, Kanagawa Prefecture; Research Center for the Arts and Arts Administration, Keio University; Sendai City Museum; Taro Okamoto Memorial Foundation for Contemporary Art; Taro Okamoto Museum of Art, Kawasaki; Tendo Co., Ltd.; Kazuko Usui; Galerie Yoshii, Tokyo; and Shigeru Yura.

For those in the United States and other countries who have provided loans of objects, permission to reproduce photographs, and necessary research we thank you too: Ordway and Jean Burden; Center for Creative Photography; The Museum of Modern Art, New York; Bruce Nichols; Isamu Noguchi Foundation and Garden Museum, New York; Andrew Popper; Shoji Sadao; and Eiko Sakai. Thanks are also extended to Stephanie Salomon and Garrett White, without whom there would be no exhibition catalogue.

The leadership of Samuel Sachs II, Chairman of The Noguchi Museum, and the ongoing support of his fellow trustees, Donald Elliott, Frances Daly Fergusson, Christiane Fischer, Hugh Hardy, David Holbrook, Peter Jachym, Lucy Lamphere, Richard Lanier, Martin Margulies, Thomas Messer, Malcolm Nolen, Isaac Shapiro, Yoshio Taniguchi, and Seiji Tsutsumi, as well as of our esteemed Honorary Life Trustees Priscilla Morgan and Shoji Sadao, must also be acknowledged. Their shared spirit, invaluable contacts, and unwavering commitment enabled the museum staff to realize this unprecedented project.

Our colleagues at the Isamu Noguchi Foundation–Japan have also been of great assistance. We thank each of you for your help on all our travels to Japan and for your support of *Design: Isamu Noguchi and Isamu Kenmochi*.

Last, but by no means least, The Noguchi Museum in New York has a terrific staff, comprising a most effective team, as witnessed in the creation of this exhibition. Within the curatorial department, Bonnie Rychlak's vision has been supported by Heidi Coleman, Larry Giacoletti, and Carl Riddle. Museum research assistant Nina Murayama, a doctoral candidate in art history at The Graduate Center, The City University of New York, has worked tirelessly on the exhibition and provided critical Japanese-English translation assistance. Heather Brady, Jennifer Burlenski, Horacio Costano, Peter Daly, Serena D'Arcangelo, Douglas DeNicola, Christine Dillon, David DiVincenzo, Amy Hau, George Juergens, Scott Malbaurn, and Peter Scibetta have all contributed significantly through their ongoing efforts and continued good work.

It is our hope that the outcome of this collective effort will interest and inform a wide audience on a previously unexplored aspect of mid-twentieth-century design and these two masters of the genre. Although Kenmochi had been influenced by the West and studied with major European designers, Noguchi was the first American he worked with. Their collaboration and their continued sharing of ideas resulted in something singular. Just as Noguchi has come to be revered in the international design community, we hope that a new focus will be brought to bear on his friend and colleague, the other Isamu.

Jenny Dixon
Director, The Noguchi Museum

8/9

PREFACE

Design: Isamu Noguchi and Isamu Kenmochi is a rare instance of a cross-cultural design exhibition. The show sheds light on the historic development of industrial design in Japan, and its relation to Europe and the United States.

Isamu Noguchi worked in many artistic fields. Industrial design was one of the areas in which he left a legacy that has come to enjoy iconic status. Never aligning himself with any professional group propounding one theory or another, his works spoke for themselves; they were mass-produced by companies such as Knoll Associates or the Herman Miller Co. in the U.S., or handmade in home workshops, such as the Akari LIght Sculptures from Gifu, Japan. Their success in design and sales is partially due to Noguchi's advantageous use of his roots and relations on both sides of the Pacific Ocean.

Isamu Kenmochi worked in Japan alone, but there he too found great success. He was trained in the spirit of early European Modernism and matured as an alert witness to postwar American Modern. His work had decisive effects on the development of industrial design in postwar Japan. He played a role within the institutions of the Industrial Arts Research Institute (IARI) and the Japan Industrial Designers' Association (JIDA), both governmental organizations for the promotion of modern design in Japan; this official role gave him a voice that was heard at institutional levels. Additionally, his creations for his own design firm became well known to the public. Whether as an agent for design institutions or as a designer, Kenmochi's ideas grew from both his native Japanese culture and overseas influences alike.

Both designers, then, incorporated ideas from Western and Eastern cultures; yet, each differed in personality, life choices, and the modeling of their careers. Noguchi, the highly individualistic, restless wanderer between cultures, disciplines, and countries; Kenmochi, the locally grounded searcher for bridges between tradition and modernity, and between collective and individual production. Even so, they met, influenced, and befriended each other. This fascinating relationship is the topic of the current exhibition.

The Noguchi Museum is the perfect place to realize such an undertaking. The Museum is firmly rooted in the two cultures of West and East, maintaining museums in the U.S. and in Japan. Both facilities recently underwent renovations that transformed them into state-of-the-art museums. However, they have retained the original spirit that they possessed when they were simply Noguchi's studios. The museum in Long Island City in particular has made the transition from a place that only preserves a legacy to one that embraces both preservation and innovation.

In celebration of the centennial of Noguchi's birth in 2004, the just-reopened museum featured a new installation of the permanent collection—the first reinstallation since Noguchi's passing away—as well as its first temporary exhibition. This show, *Isamu Noguchi: Sculptural Design*, happened to focus entirely on Noguchi's design works. Since this initial exhibition the museum has excelled in rotating shows. The variety of thematic approaches exemplifies a wide range of viewpoints concerning Noguchi's significance in the world of art and design. *Design: Isamu Noguchi and Isamu Kenmochi* is a valuable addition in this line.

This catalogue gathers together an impressive collection of writings about the relationship between the two designers and their influence upon each other. It also includes historic publications written by Noguchi and

Kenmochi themselves. Bonnie Rychlak, curator at The Noguchi Museum, introduces the theme of the exhibition with a historical overview of the way in which each absorbed foreign influences into his work: Noguchi by actual exposure to life, art, and culture in Japan, the U.S., and Europe; Kenmochi mostly by exposure to ideas from modernist Europe and America that were imported to Japan by foreign experts and designers. Their encounter thus was an interesting dialogue of two artistic minds that applied similar influences in different ways. Noguchi's approach was direct, intuitively integrating those influences into his artistic work, whereas Kenmochi was more intellectual and abstract in his application of those influences to his designs.

Hitoshi Mori, an expert in Japanese design history and its relations to America and Europe, discusses several design objects by Noguchi and Kenmochi that show traces of both Western and Japanese tradition and culture. However, the mixing of domestic and foreign features was an idea not easily acceptable to the Japanese public of the postwar years. On the one hand, there was critical disapproval of foreigners who designed in a Japanese mode (Noguchi, being born in Los Angeles and having spent the majority of his life in America, was considered a foreigner in Japan); on the other hand, much praise was heaped upon the same foreign designers and experts for their rediscovery of Japanese traditional design principles. At the time, "Good Design" in Japan was supposed to be in Western style. This paradoxical situation between a reappraisal of "Japan-ness" and a fancy for Western style affected not only Noguchi as a "foreigner with a Japanese spirit" but also Kenmochi as a "Japanese with a foreign spirit."

Tetsuo Matsumoto provides the reader with an insight into the difficulties that Westernization brought upon Japanese individuals in terms of their personal roles in a new industry. He discusses Kenmochi in the double-bind of his design process. Having worked with European designers in Japan, and having met American designers at home and overseas, Kenmochi was acquainted with two opposing principles of work and production: the Japanese model of collective organization and consensus versus the Western idea of individually driven single-handed creation. "The nail that sticks out gets hammered down" is synonymous with the rigorous Japanese expectation of the individual in relation to the group, leaving little space for individual excellence. As a result, Kenmochi led a lifelong struggle between a designer's need for personal expressivity and the need for official integration into his society.

Nina Murayama retraces Noguchi's design in an article that focuses on the Akari Light Sculptures. Akari are a symbol of the marriage between Japanese and American ideas. Based on traditional Japanese material and handcraft skills, the lamp shades captured a romantic exoticism that gave them appeal to a Western sensibility as functional objects, and to the Japanese as both modern illumination and artworks. They were instantly successful on the American as well as the Japanese markets, and this successful and long-lived design project germinated during the time of Noguchi's and Kenmochi's first meeting.

In sum, *Design: Isamu Noguchi and Isamu Kenmochi* is an important contribution to the understanding of cross-cultural influences upon the development of both designers' work. Undoubtedly, their products deserve their firm places in the history of design as significant and original pieces. The traces of each man's struggle with cross-cultural influences are conceptually representative of a whole generation of designers in Japan. Thus, the encounter of Noguchi and Kenmochi offers invaluable insight into the larger realm of conceptualizing design with the Different, the Foreign, the Other.

Katarina V. Posch
Associate Professor in Design History and Theory, Pratt Institute, New York

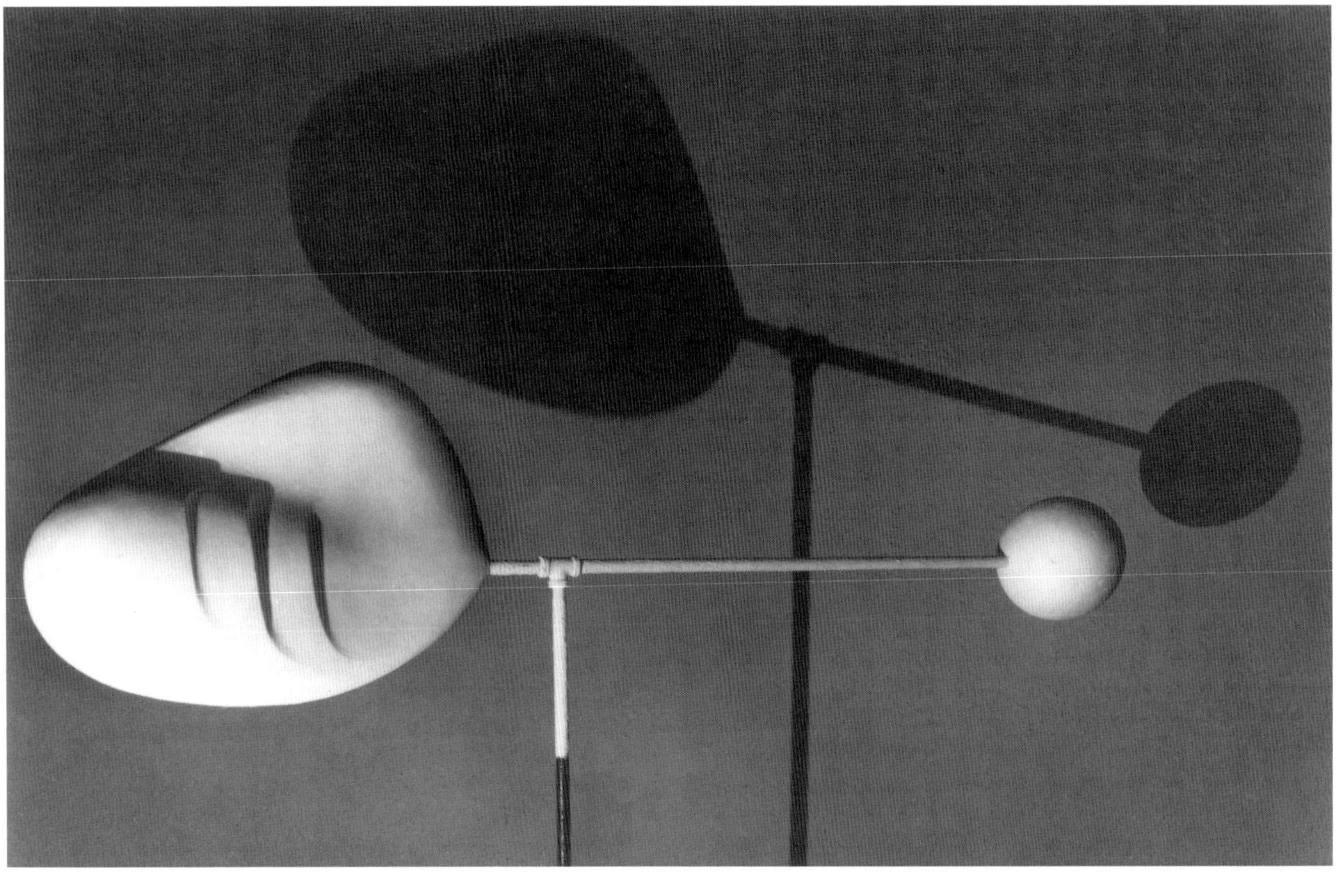

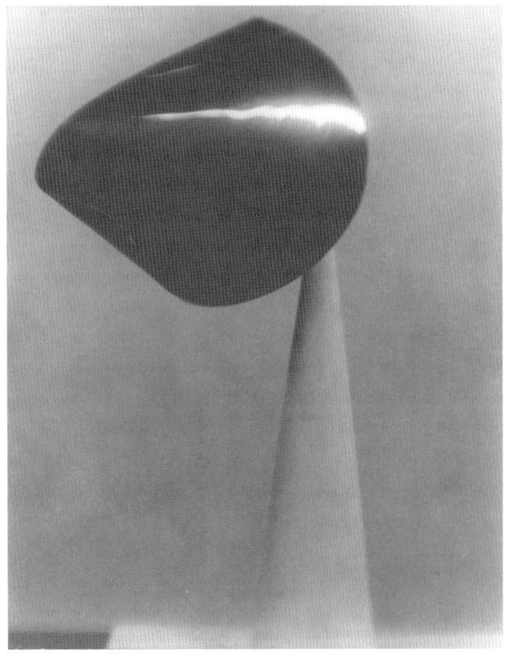

Top: Isamu Noguchi, Model of Musical Weathervane, magnesite, 1933, dimensions not known. Bottom: Weathervane, magnesite, 1928, dimensions not known. Opposite: Musical Weathervane (Lunar Weathervane), pencil on paper, 1933; dimensions, clockwise from top left: 19 3/4 x 12 3/4 in.; 19 3/4 x 12 3/4 in.; 19 3/4 x 12 3/4 in.; 25 1/2 x 19 3/4 in. One of Noguchi's earliest designs, the Musical Weathervane used a simple fluting device to catch the wind and create a whistling sound. If produced, it would have included wiring for nighttime illumination.

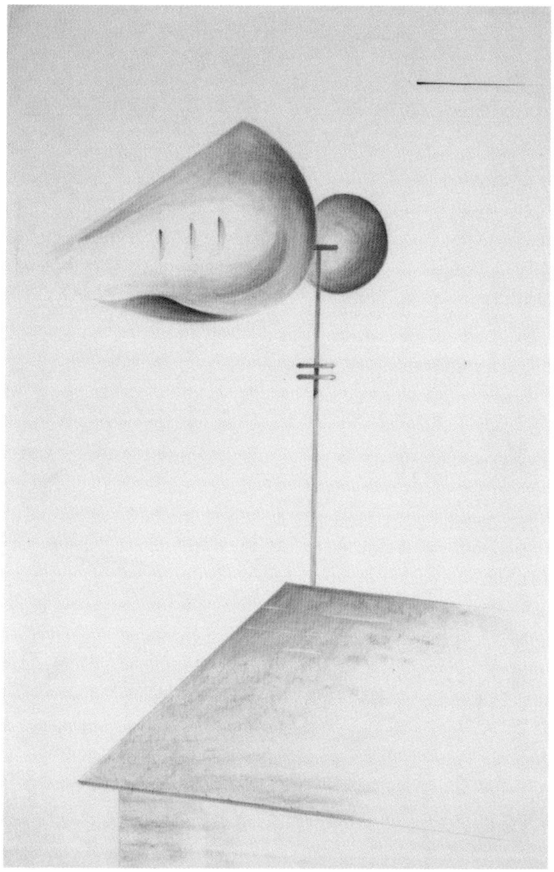

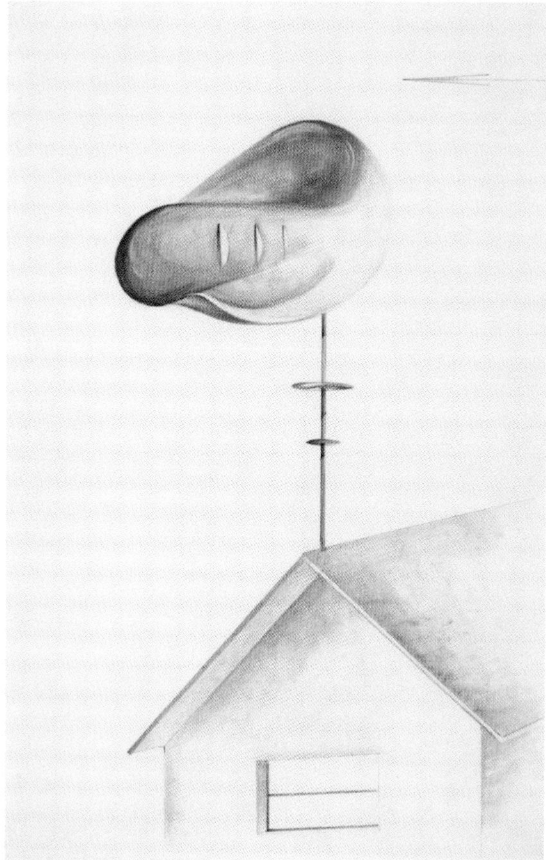

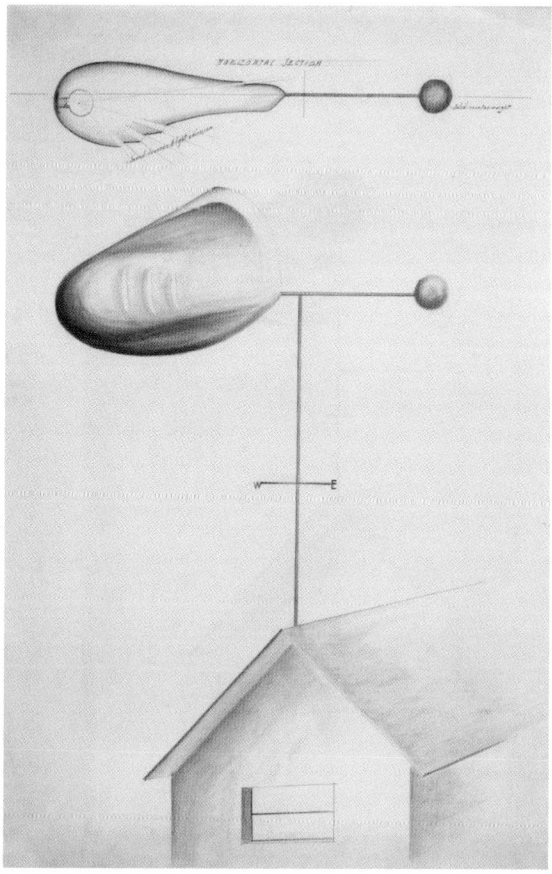
HORIZONTAL SECTION
W
E

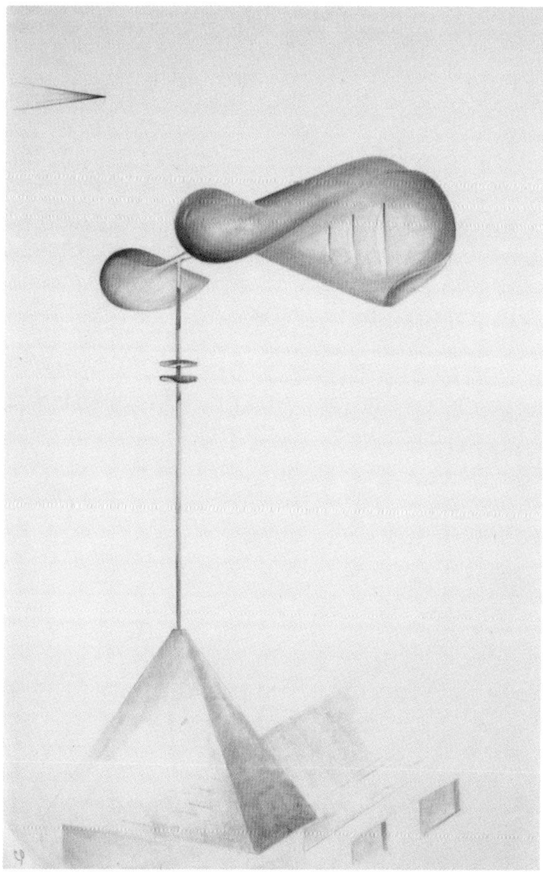

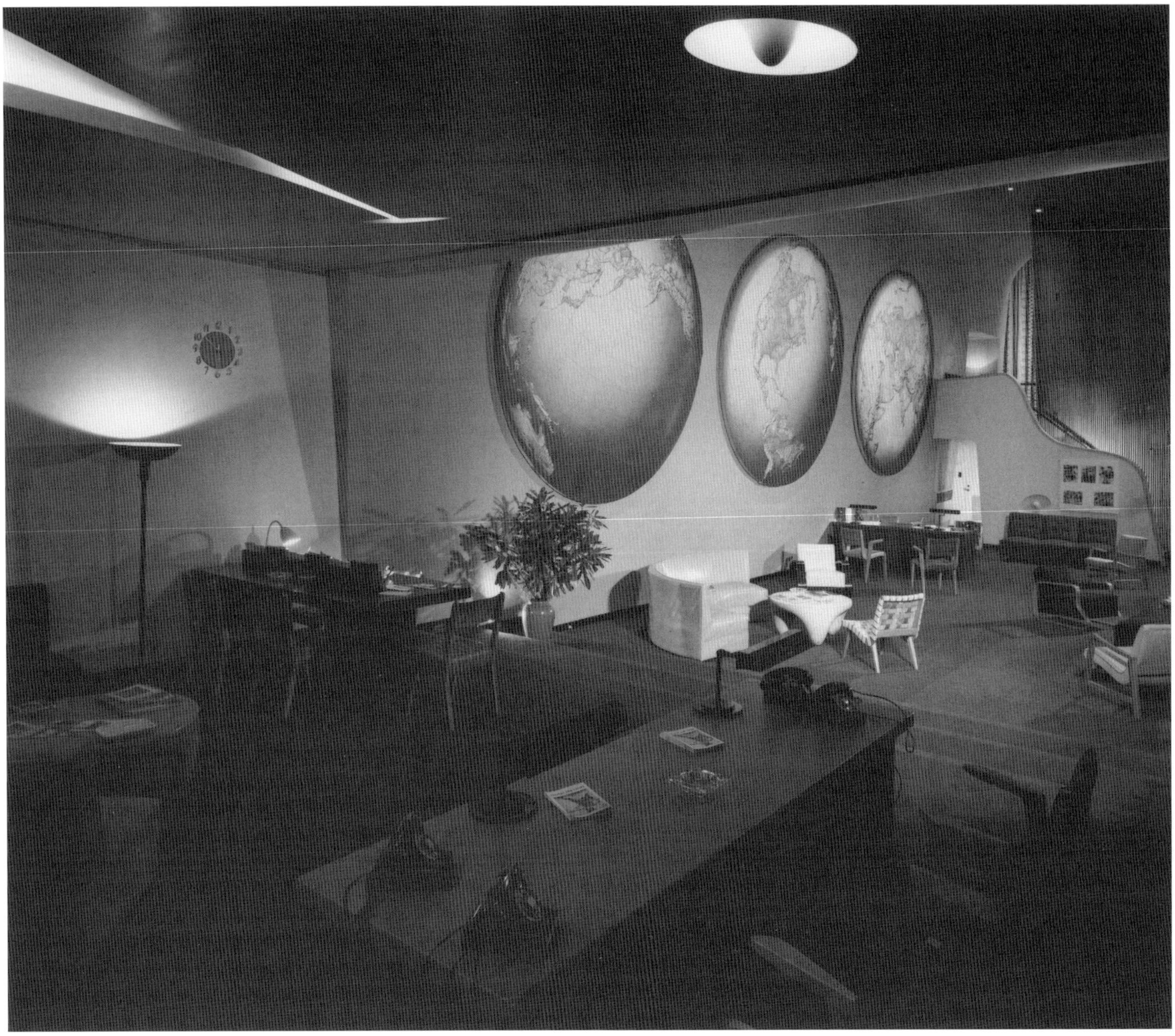

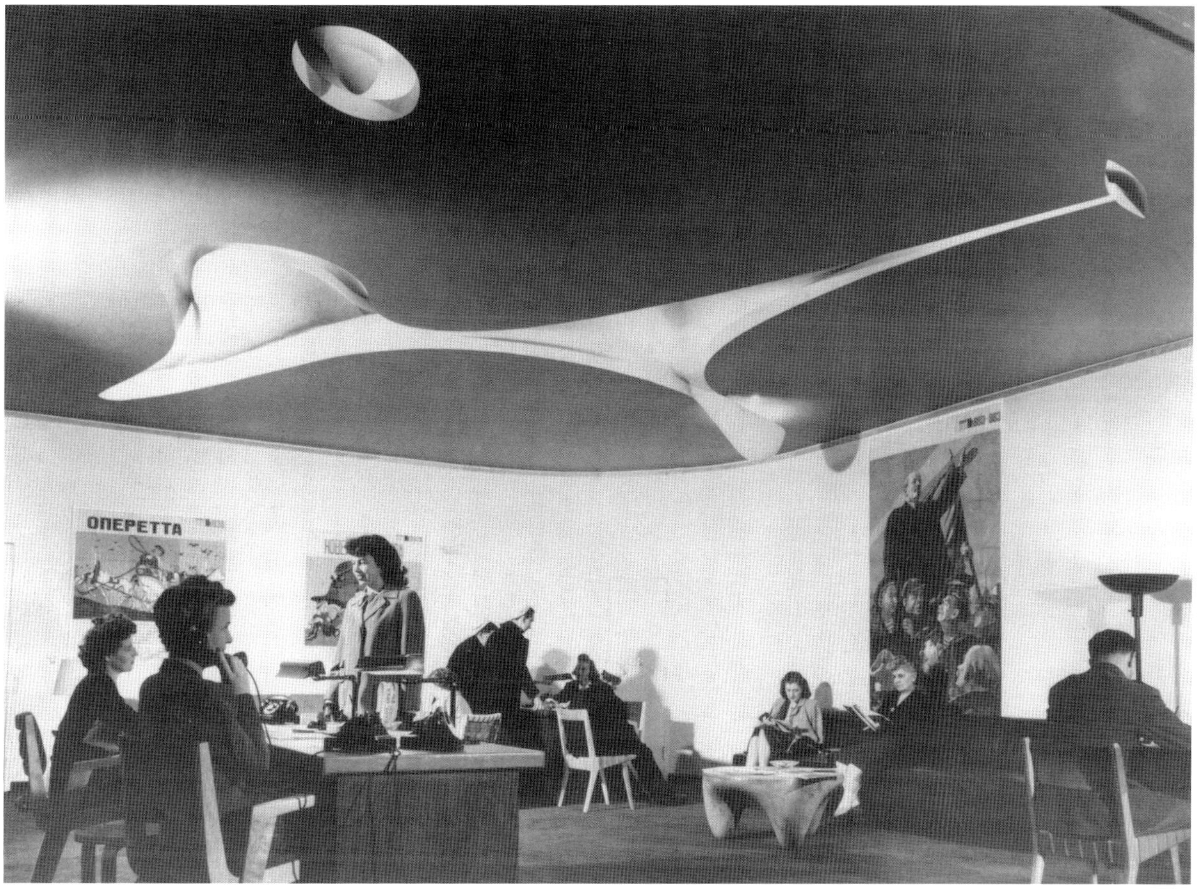

Above and opposite: Isamu Noguchi, Interior of the Time and Life Building, New York, 1947. The original home of Time, Inc., was the 1937 Time and Life Building at 1 Rockefeller Plaza by architects Harrison and Abramovitz. Ten years after the building was completed, Noguchi was commissioned to add to the interior spaces of the lobby and reception area. Along with architect Wallace K. Harrison, Noguchi designed an illuminated, free-form ceiling above the reception desk, along with other items, including tables and a wall relief. Noguchi's alterations were lost when Time, Inc., moved to the new Time Life Building, 1271 Sixth Avenue, in 1959.

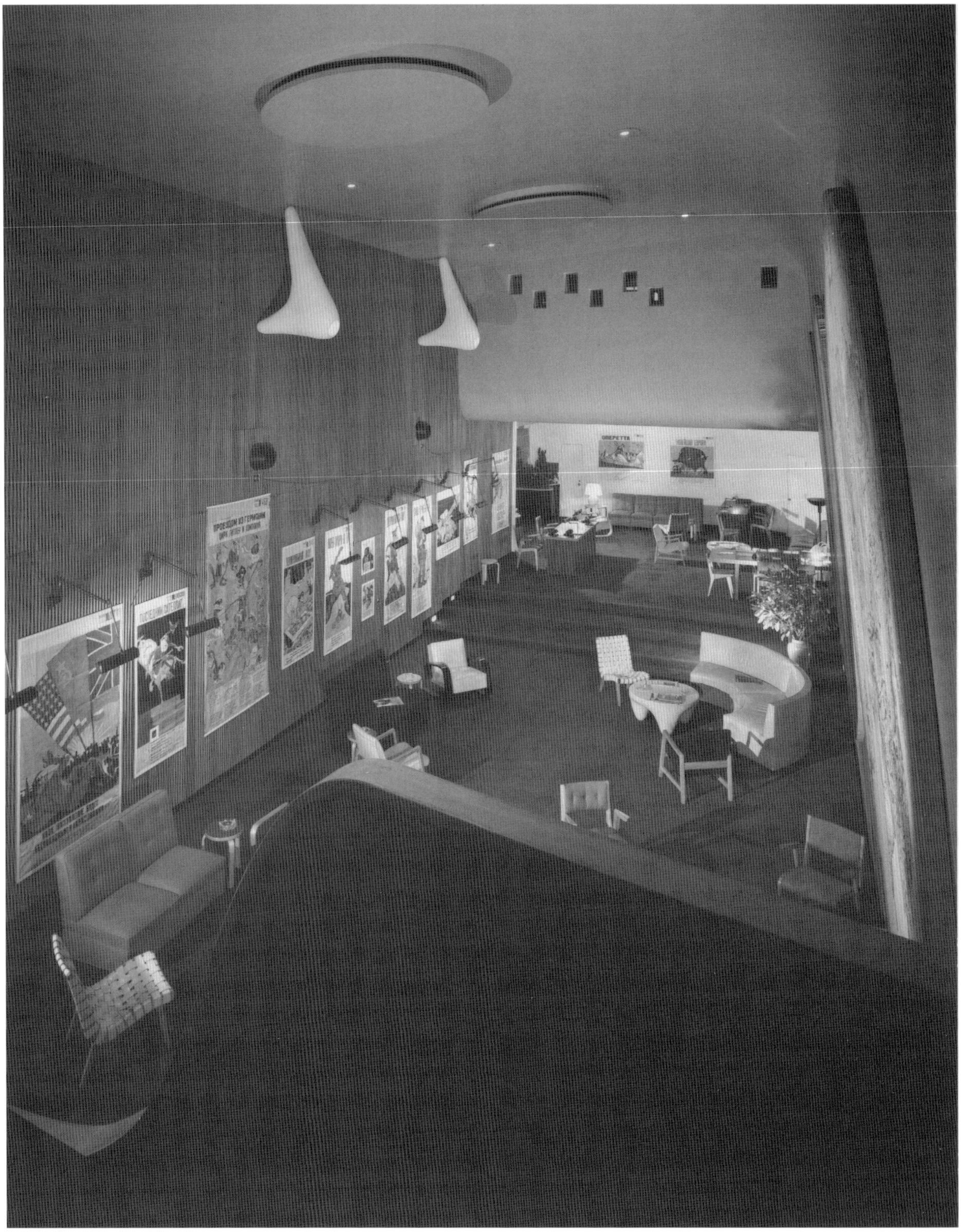

Isamu Noguchi, Interior of the Time and Life Building, 1 Rockefeller Plaza, New York, 1947.

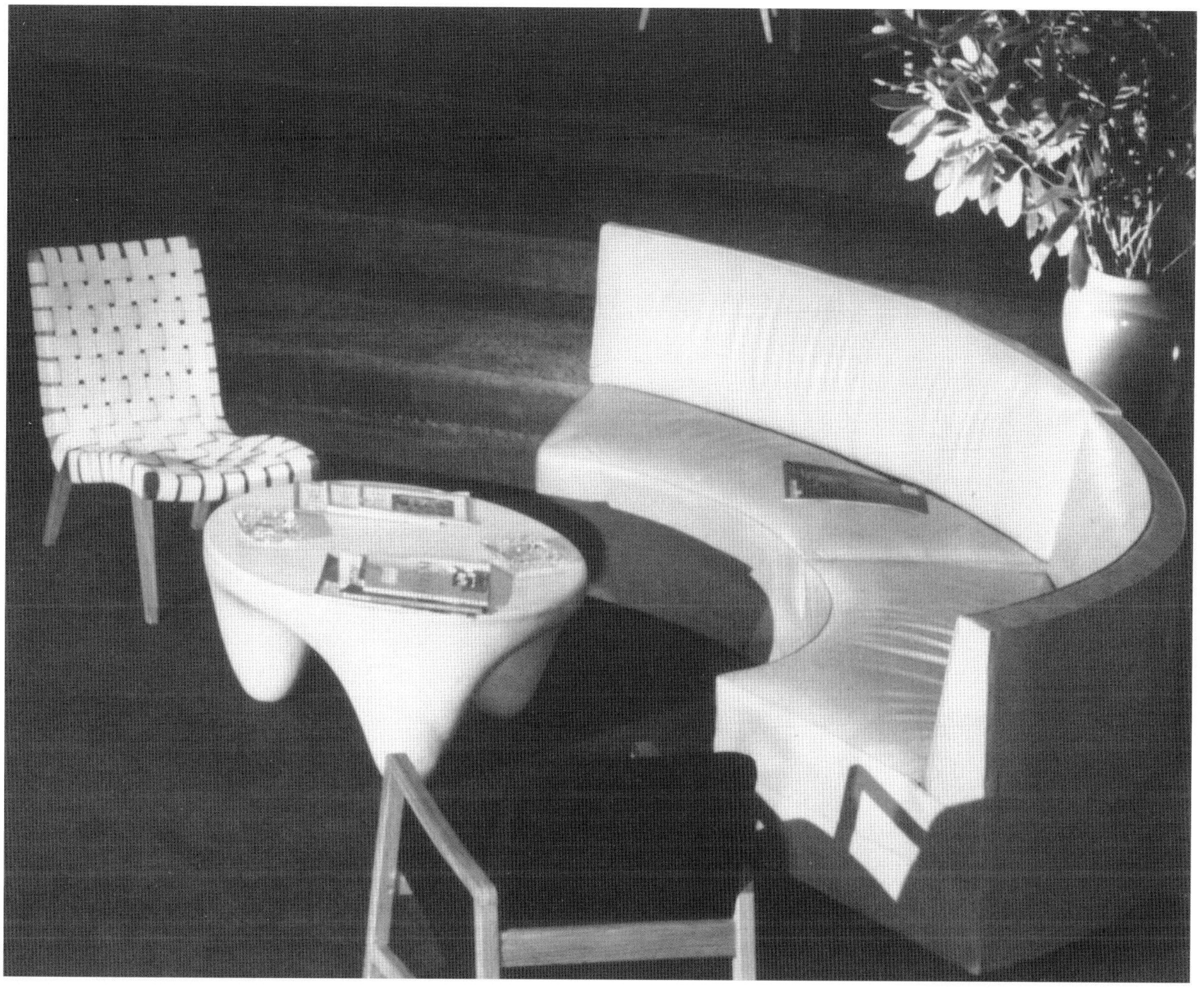

Isamu Noguchi, Table for the Time and Life Building, 1 Rockefeller Plaza, New York, 1947.

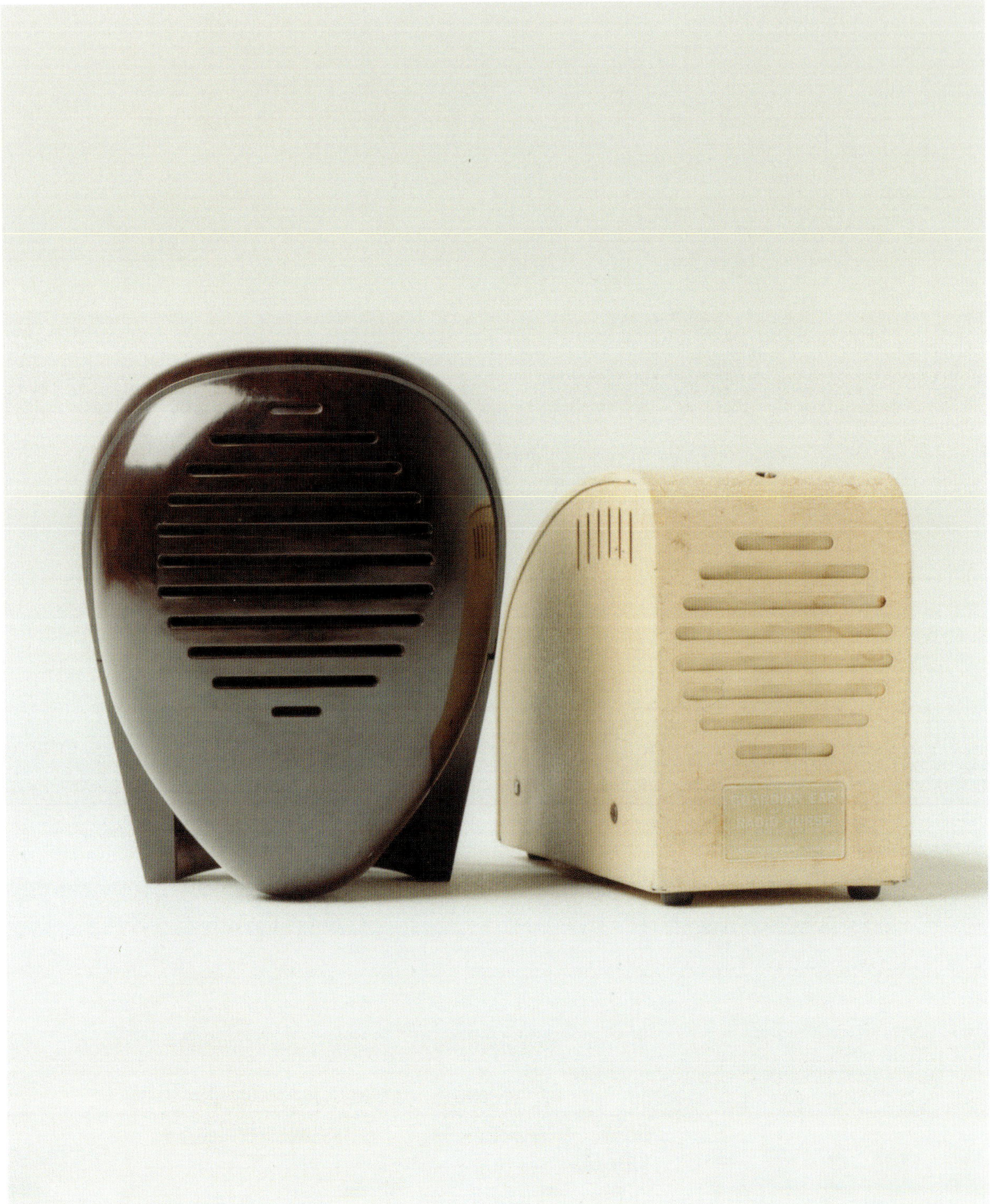

Above left: Isamu Noguchi, Radio Nurse, Bakelite, 1937, 8 1/4 x 6 1/2 x 6 in; above right: Guardian Ear, metal, 1937, 6 1/2 x 6 x 9 in. Noguchi's first mass-produced object, the child-monitoring device Radio Nurse and Guardian Ear was commissioned by Zenith Corporation. The sleekly designed Radio Nurse, from the new plastic material called Bakelite, contained the speaker, and the companion piece, Guardian Ear, held the microphone for the child's room. Opposite: Radio Nurse, 1937.

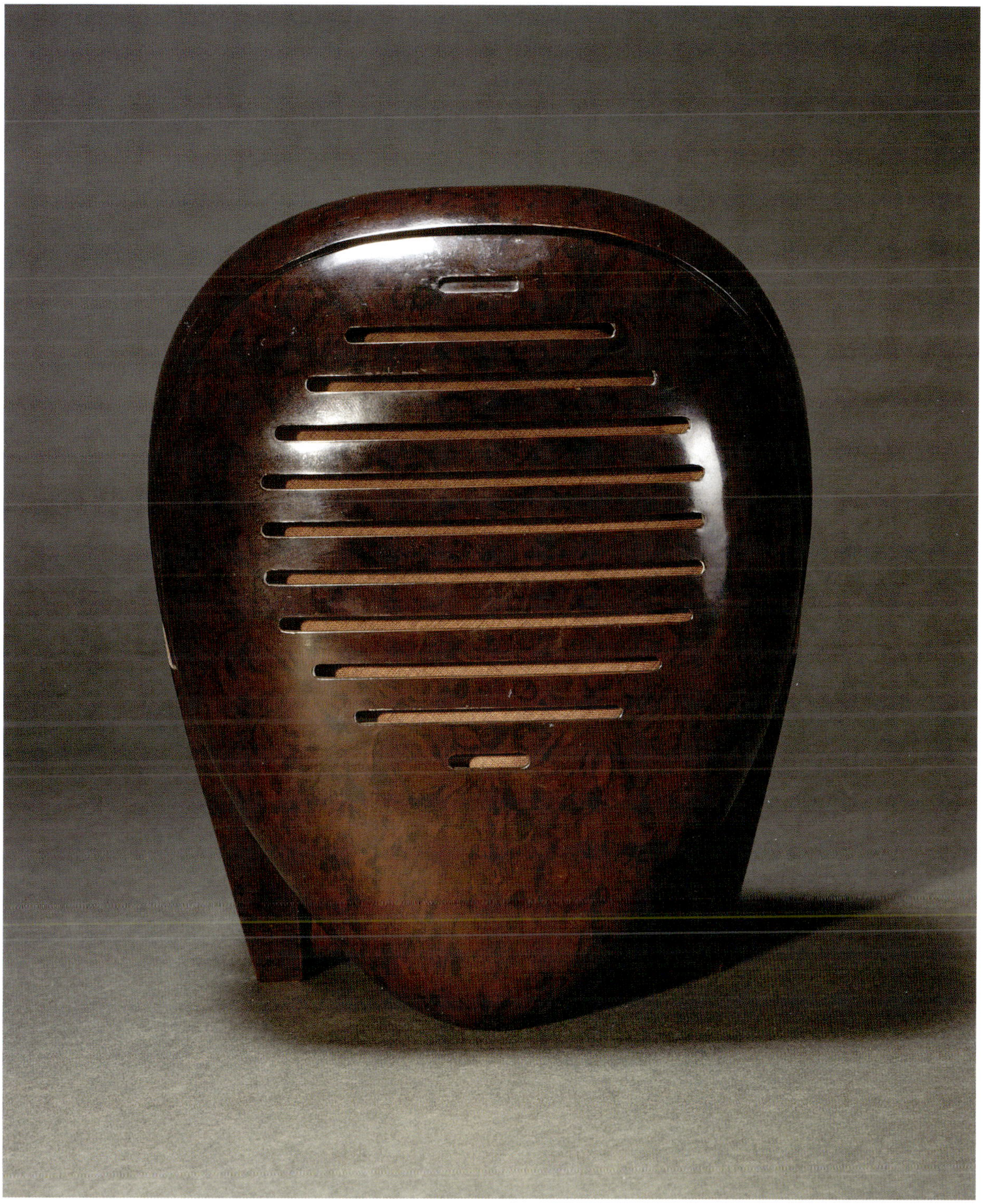

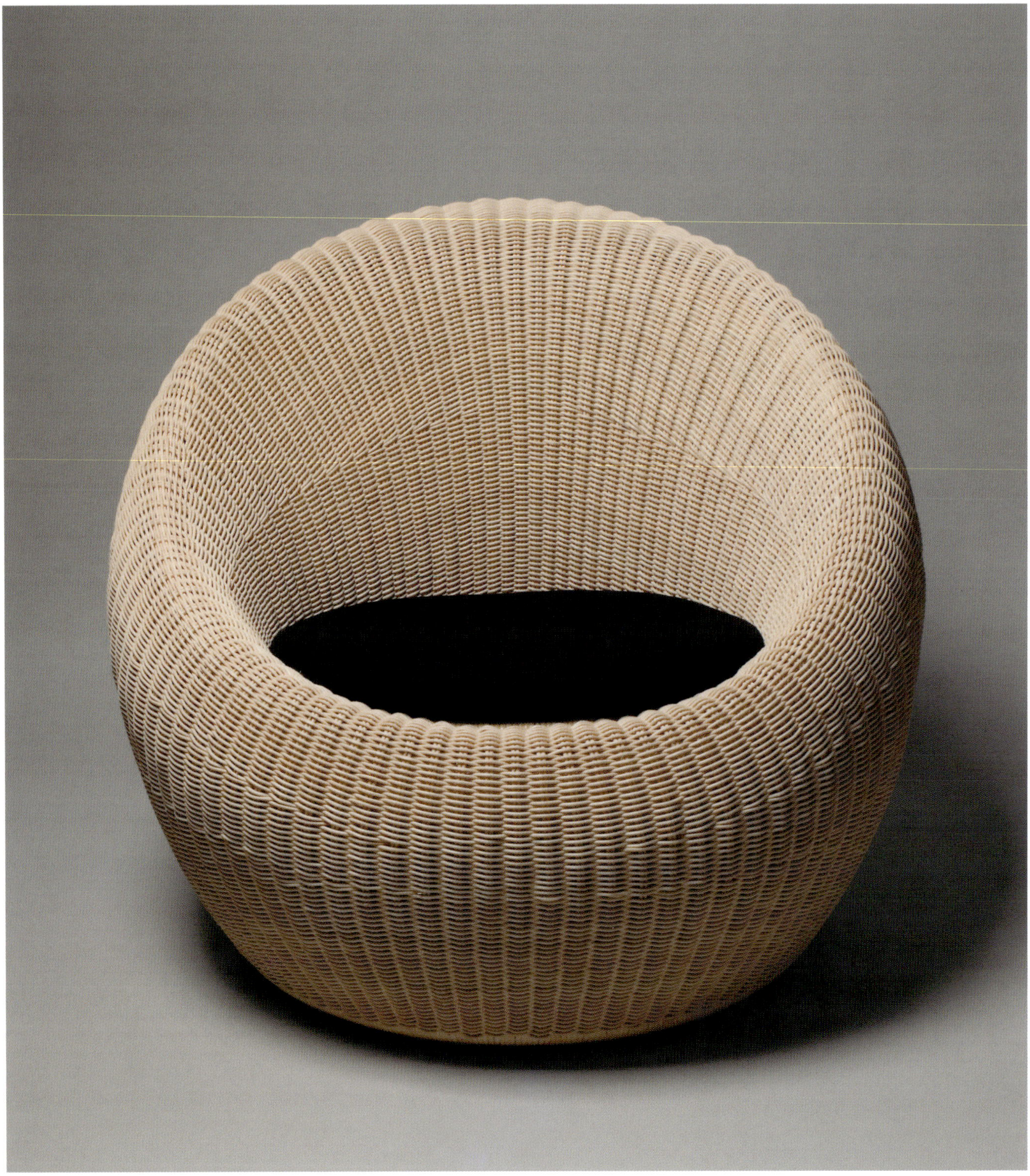

Kenmochi Design Associates, Rattan Round Chair C-3150, 1960, 27 9/16 x 31 15/16 x 27 9/16 in.
Manufactured by Yamakawa Rattan (current reproduction by Y.M.K. Co., Ltd.).

SEARCHING FOR THE NEW AND AUTHENTIC:

SHAPING THE DESIGN PRODUCTION OF ISAMU NOGUCHI AND ISAMU KENMOCHI

BONNIE RYCHLAK

Whenever we encounter some truly novel object, one that reinvents the margins of our world, we frequently ask ourselves if we have seen it before—whether it is new, old, or simply a piece of manufactured nostalgia. These questions or sensations are, more often than not, the consequence of overexposure, of too much abbreviated information that clutters our cultural experiences, making them seem unauthentic, unoriginal, or banal. Moreover, artistic artifacts have been so deeply and thoroughly mined of meaning that what might appear fresh or unexplored dimensions of history are habitually sidelined and critiqued as simply clichéd and passé. These are the narrow limits of our postmodern existence overtaking us, conditions that shape our experience of what is genuine, and of progress, and how we see the advancement of inventions. Such is the case with many of the iconic designs produced by Isamu Noguchi (1904–1988) in the United States and Isamu Kenmochi (1912–1971) in Japan.

In America and Japan, the innovative design credits of Isamu Noguchi are easily recognized, if not by name, then most certainly by form, particularly with the surge in interest in 1950s modern design occurring in the last two decades, which has given Noguchi's furniture production a rebirth of sorts. On the other hand, the work of Isamu Kenmochi, while not so well known in the West, is popularly recognized in Japan. Eight years younger than Noguchi, Kenmochi lived his entire life in Japan, whereas Noguchi, though he was raised in Japan from age two until he was thirteen, was American and spent most of his life in the United States and Europe. Kenmochi made his first trip abroad to the United States in 1952 at the age of forty.

The only work by Kenmochi to have garnered recognition in the United States is his Rattan Round Chair from 1960. Selected for inclusion in the permanent collection of The Museum of Modern Art in 1964, it is more widely known through unauthorized mass-market reproductions sold at inexpensive Asian emporiums. The Rattan Round Chair in its original form embodies Kenmochi's design principles, illustrating the perfect match of machine and hand, the human and the natural, reason and emotion—contradictory terms and attributes that eluded Kenmochi until the creation of this Rattan Round Chair. The natural warmth and softness of rattan is translated into a stable construction that enwraps the sitter, its simplicity of form eschewing all decoration.[1] The chair's design embodied what Kenmochi would come to describe as "Japanese Modern."

Isamu Kenmochi was one of Japan's premier postwar industrial designers, inventing modern objects suffused with a traditional Japanese sensibility. Chair design occupied a special position within his oeuvre. He was particularly interested in remaking the Western chair for daily use in Japan. Chairs were first used in Japan in public spaces and entered the private realm by way of established Western institutions and commercial quarters during the 1920s, when they began to replace the traditional mode of seating—tatami mats placed on the floor—and symbolized the Westernization of daily life.

Kenmochi created his own designs for chairs, such as those he made in bentwood. Although bentwood technique is old and common, the Armchair 207 from 1960 is original in its adjustment to the postwar Japanese spatial and economical context, modest in size for the new apartment buildings, and affordable as new furniture. The development of Kenmochi's bentwood furniture reflects a transformation of Japanese economic recovery to

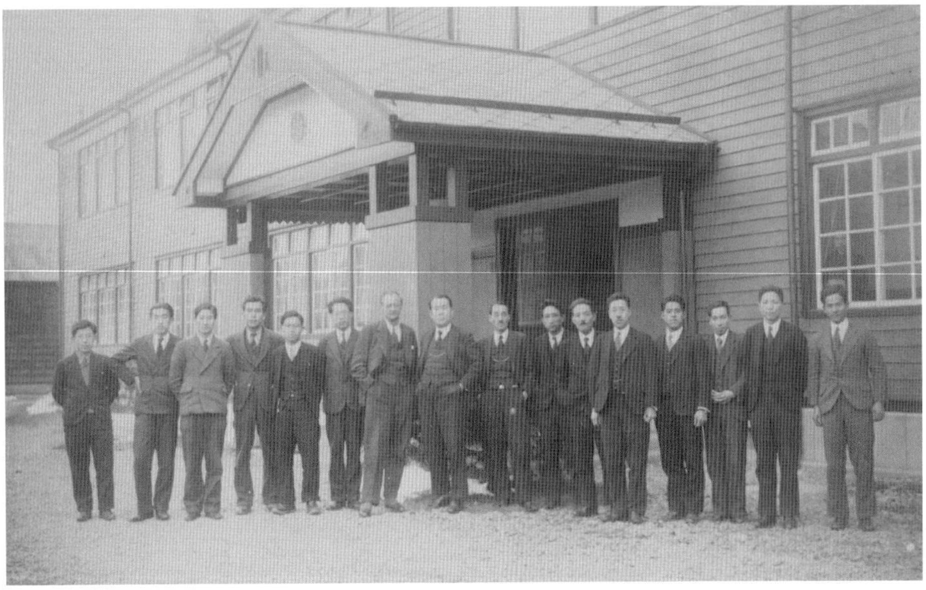

Bruno Taut (seventh from left), Isamu Kenmochi (second from left), and the National Craftsmanship Training Institute staff, 1933.

prosperity. Conscious of these conditions, he sought to balance aesthetics with the practicality of functional objects that had deep roots in his past. "I choose chairs as my first assignment," he explained, "because the chairs produced in Japan were particularly badly made."[2] This was his reason for using concave elements that followed the curvature of the body, a point of departure that illustrates just how much chair design fostered his understanding of the modern. Kenmochi wrote frequently on issues related to the chair and the standardization of its design.[3]

The direct, collaborative working relationship between Noguchi and Kenmochi was actually very brief, occurring in the two years from 1950 to 1952, and their most concentrated work together lasted only two weeks, in August 1950 when Noguchi began his fabrication of furniture and sculpture for a recent commission from Keio University in Tokyo. Yet together and individually they created a group of universally exceptional functional objects, using designs that incorporated a basic beauty of simplicity grounded in knowledge of natural materials and a shared vision that embraced experimental techniques and resources. Moreover, they remained friends until Kenmochi took his own life in 1971.[4]

The following brief exploration of the furniture production of Isamu Noguchi and Isamu Kenmochi is anchored in the simple observation that good design contributes to the advancement of more good design. A review of the influences on both Noguchi and Kenmochi is intended to confirm this optimistic conception of progress, and perhaps negate some of the postmodern tendencies toward marginalizing ideas and objects that have become overexposed in our popular cultural canon.

While examining their respective influences, it would not be fair, or indeed accurate, to grant absolute credit for originality to either of these men in their efforts to create good design. Their production was a consequence of many circumstances in addition to their shared sensibilities and an appreciation of other artists' histories and cultural and ethnic backgrounds. Noguchi, an American artist, trained with such fine-art luminaries as Gutzon Borglum and Constantin Brancusi. Kenmochi, a Japanese industrial designer who studied woodworking, architecture, and engineering, developed, under the tutelage of Bruno Taut (1880–1938), an aesthetic disposition that was equally determined. Taut, an architect and expatriate who came to Japan as a refugee from Hitler's Germany in 1933, derived his teachings from the Bauhaus, Le Corbusier, and Mies van der Rohe.[5] Of particular impact on Kenmochi were the theories espoused by the Bauhaus and its founder, Walter Gropius, especially their efforts in Germany before World War II to industrialize indigenous craft traditions.[6]

Bauhaus theories aimed to teach the arts and crafts in tandem and to bridge the widening gulf between art and industry. There was no instruction in the history of these fields at its successive schools because design and creation were intended to follow Bauhaus principles rather than precedent. One of the most important contributions of the Bauhaus was the invention of modern furniture design. To Kenmochi, the concept of merging art and industry held both aesthetic and economic significance for postwar Japan.

Long before Noguchi arrived in Japan, Taut had provided Kenmochi with his most significant insights. Kenmochi began working for the Industrial Arts Research Institute (IARI) in 1933, and on September 4, 1933, Taut visited an exhibition of experimental products produced by the IARI at the Mitsukoshi Department Store. Taut promptly criticized the products on display, declaring them to be of high quality in terms of technique and carefully made but below the standards of quality or form for modern international design.[7] Improvisational curves and the play of color were features of these products and not attributes of good design as far as Taut was concerned.

Transformed from the National Craftsmanship Training Institute (a school established in the Sendai region in 1928 that offered training in traditional craft techniques), the IARI was developed under state sponsorship with the charge of redesigning crafts for export. As unmistakably expressed by the Ministry of Commerce and Industry's official statements, the IARI's mandate was to lead and promote the latest scientific and technical advancements in industry in order to create products suitable for the export market, a market requiring well-made functional objects that are, when ordered, readily available. This was in stark contrast to Japan's traditional handmade production, which emphasized crafts and craftsmanship and reliance on small cottage industries.

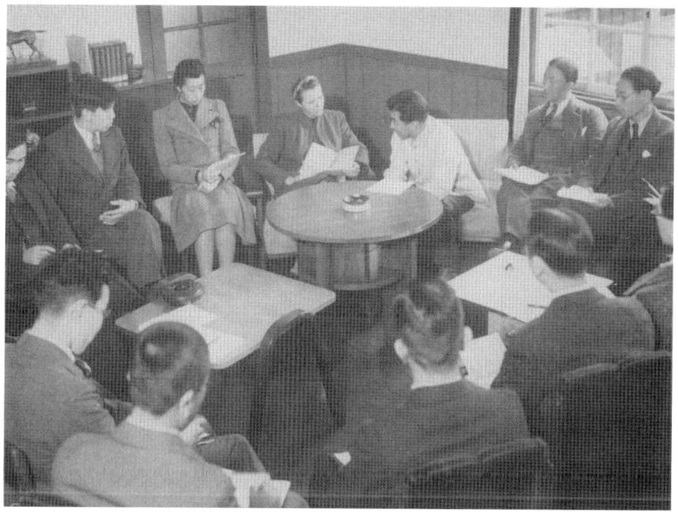

Group discussion, Industrial Arts Research Institute, 1940.
Charlotte Perriand, top center.

The IARI would determine Kenmochi's life direction, first by training him, then by launching his career as part of the state's design activities. The Ministry of Commerce and Industry, which had sponsored the IARI before the war, became the Ministry of Munitions during the imperial buildup, and Kenmochi inevitably became a munitions technician who supervised the development of wood airplanes and decoys.

The IARI was responsible for promoting exports, overcoming a depressed economy, and improving daily life. During the war, the IARI sponsored what it termed "fighting design," which entailed the construction of standardized goods for wartime life. In wartime, substitute materials were needed to keep up with military demands for natural resources. Japanese-occupied Southeast Asia was a logical source for materials such as bamboo, and the IARI investigated the use of bamboo products for daily use. The IARI also had a chemical experiment division that was responsible for developing new adhesives. This theoretical framework and influence on Kenmochi was clearly a dangerous one. Nonetheless, Japan's regime allowed Kenmochi to advance his goals for promoting the implementation of scientific methods, and for the aestheticization, standardization, and functional improvement of objects needed in daily life.[8] The Industrial Arts Research Institute was a branch of a government agency that mandated the advancement of aesthetics and technology, and advocated a new kind of functional beauty based in modern industrial technology. This included the adoption of metal structural materials and techniques, such as the application of KS magnetic steel, mixing iridescent metal powders with lacquer to produce an iridescent finish, and metal inlay methods.[9] With modern woodworking machines, education began to extend beyond

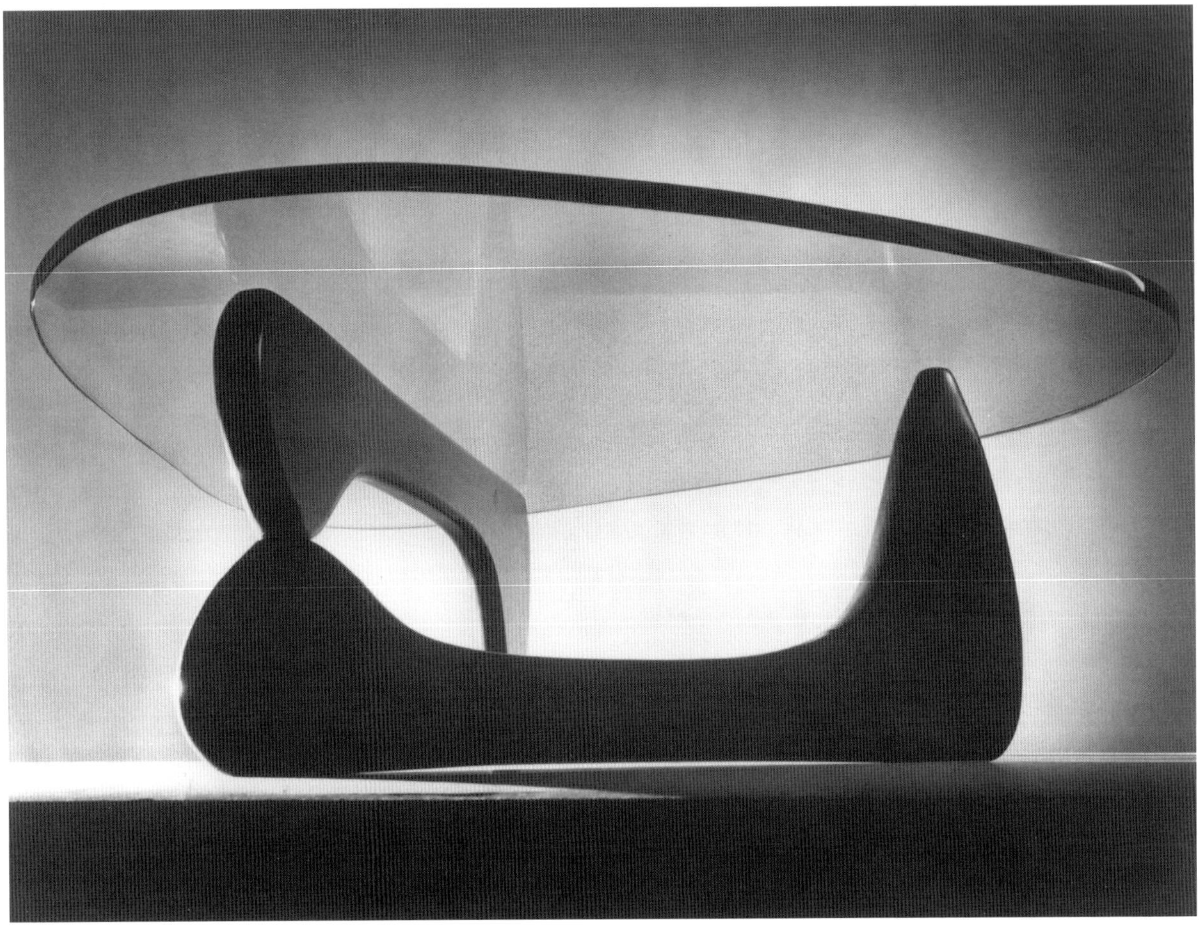

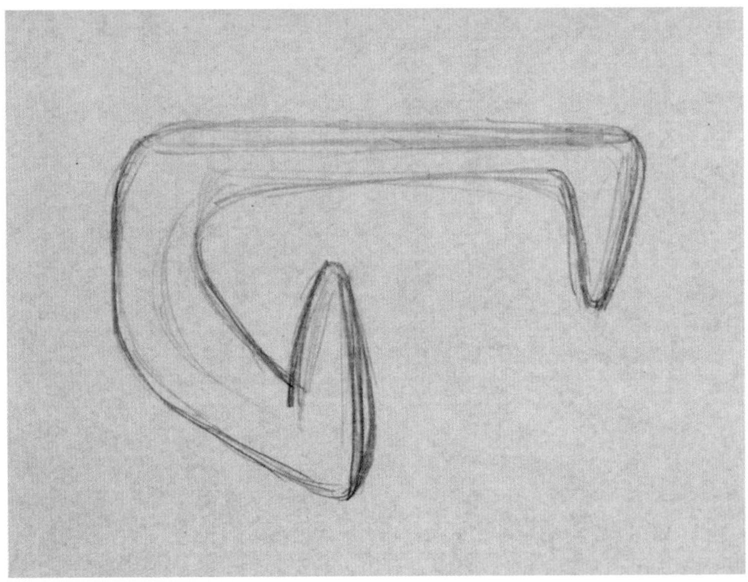

Top: Isamu Noguchi, Glass-Topped Coffee Table, wood and glass, 1944, 15 x 50 x 36 in. Bottom: Isamu Noguchi, Drawing for Coffee Table, pencil on paper, 1942–1948, 8 9/16 x 11 in. An icon of mid-century modernism, the Glass-Topped Coffee Table remains a signature design by Noguchi, embodying a balance of sculptural form and everyday function.

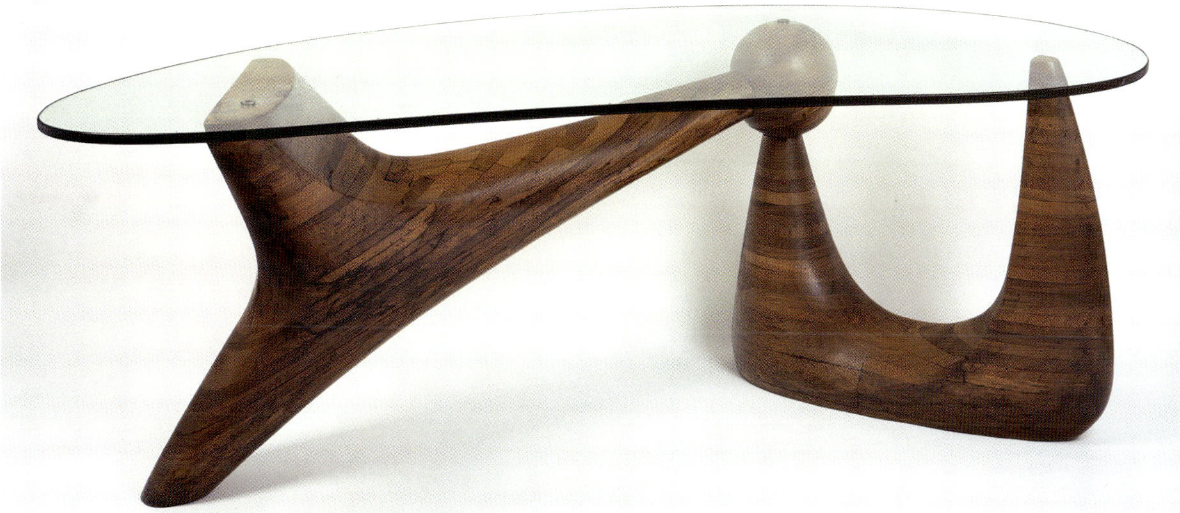

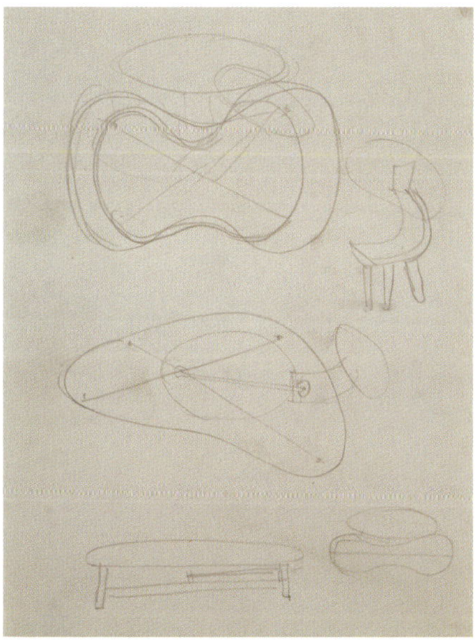

Top: Isamu Noguchi, Table, laminated rosewood and glass, 1939, 29 x 82 x 32 1/2 in. Noguchi's earliest table design developed from his experience with carving wood and stone, and from his growing interest in biomorphic forms. He received the commission for this table from A. Conger Goodyear, then-president of The Museum of Modern Art, New York. This seminal work became the conceptual foundation for his subsequent furniture designs, as well as for sculpture and set designs for dancer-choreographer Martha Graham. Bottom: Drawing, Five Ideas for Sculpture and Furniture, pencil on paper, ca. 1942–1948, 11 x 8 1/2 in.

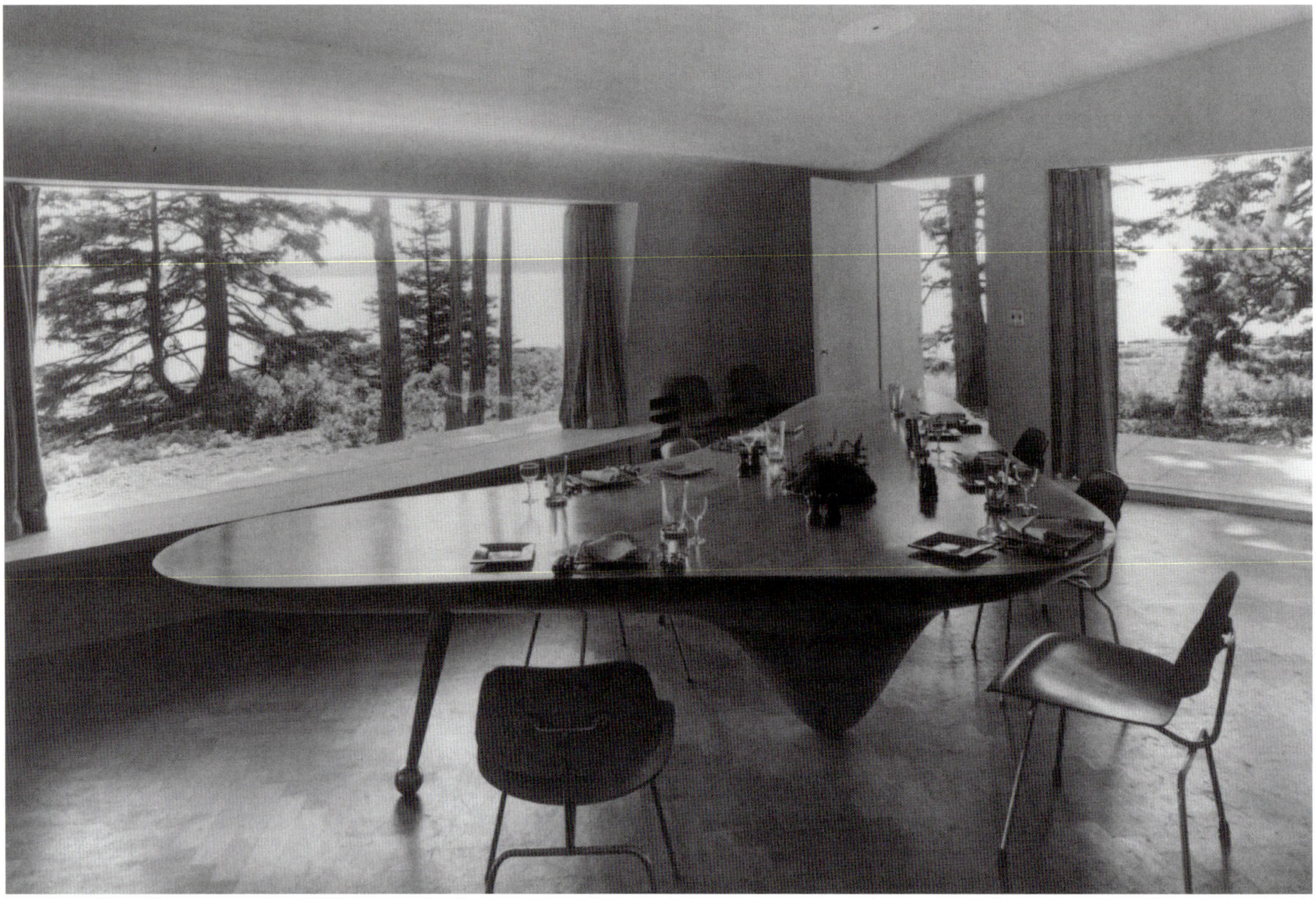

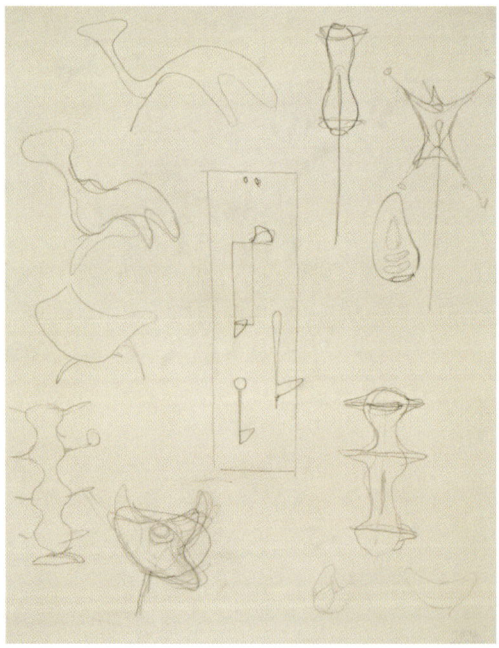

Top: Isamu Noguchi, William Burden Table, ca. 1947–1948. Noguchi designed this laminated beechwood dining table, his largest piece of furniture, as a commission for William A. M. Burden, a Museum of Modern Art trustee, in 1946. Designed in a reverse curve to the shoreline visible from the Burden summer house in Northeast Harbor, Maine, the table was lost when the house was destroyed by fire in 1999. Bottom: Twelve Ideas for Sculpture and Furniture, pencil on paper, 1943, 10 15/16 x 8 1/2 in.

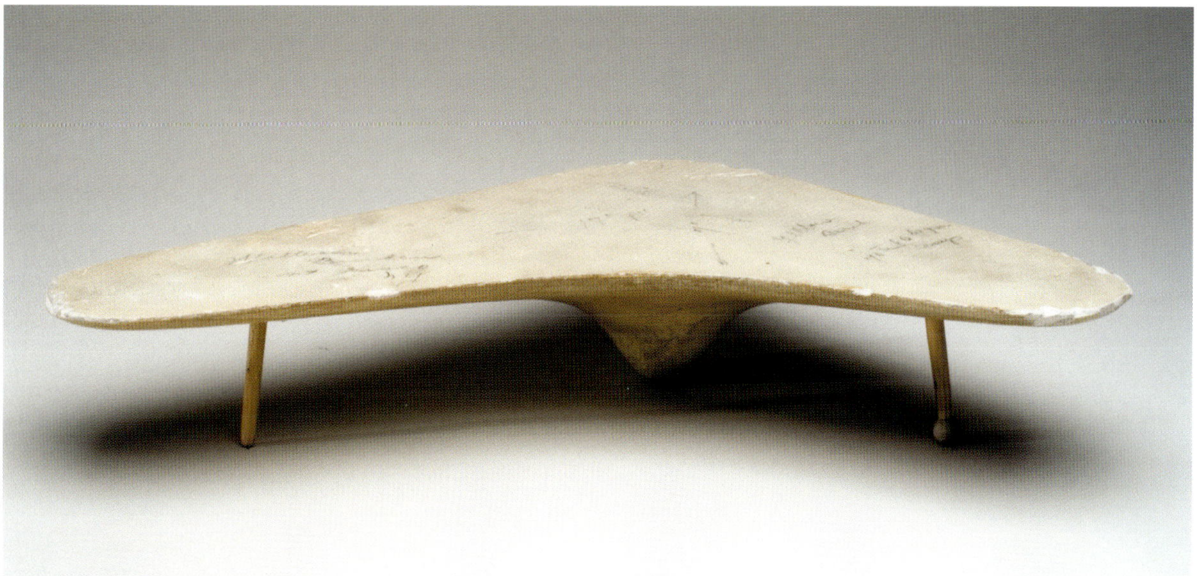

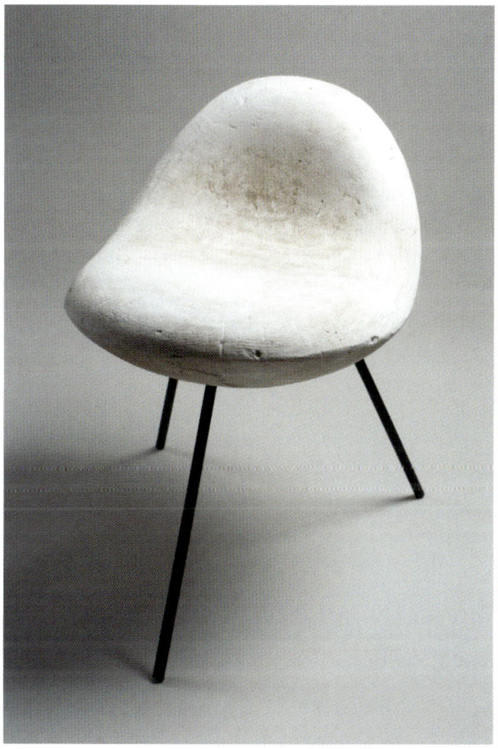

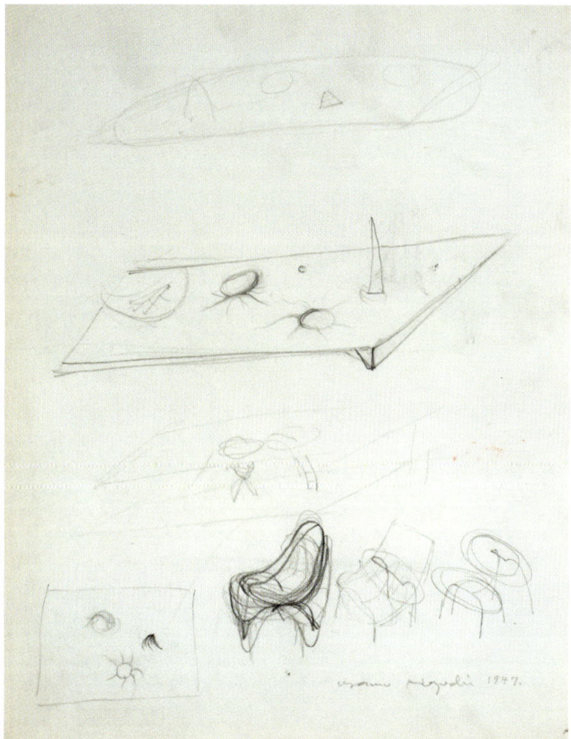

Clockwise from top: Isamu Noguchi, Model for William Burden Table, plaster, wood, and steel, ca. 1947–1948, 2 1/2 x 4 x 8 in.; Table and Chairs, pencil on paper, 1947, 13 7/8 x 10 7/8 in.; Model for William Burden Chair, plaster, ca. 1947–1948 (steel legs reproduced in 2007), 28 1/2 x 20 1/2 x 28 1/2 in.

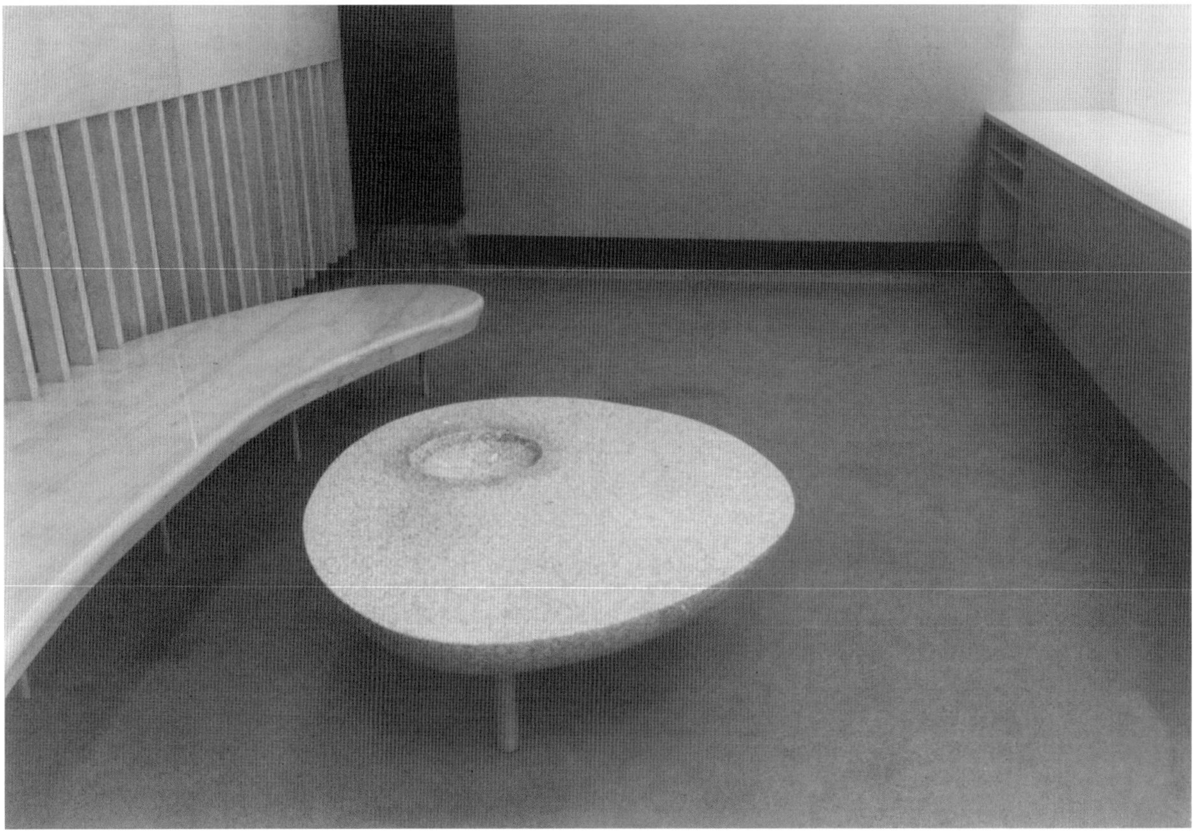

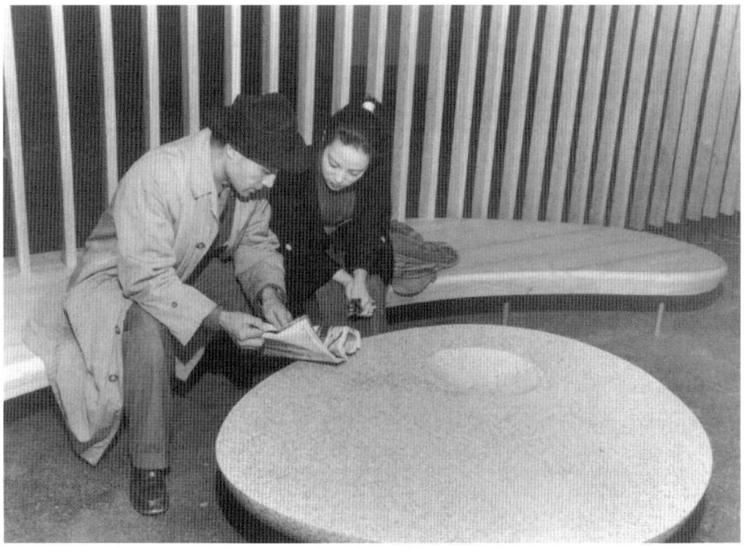

Top: Isamu Noguchi, Installation View, Table and Bench for Chuo Koron Gallery, Tokyo, granite, iron, and wood, 1952, 30 x 53 x 43 in.
Bottom: Isamu Noguchi and Yoshiko (Shirley) Yamaguchi at Chuo Koron Gallery, ca. 1952–1953.

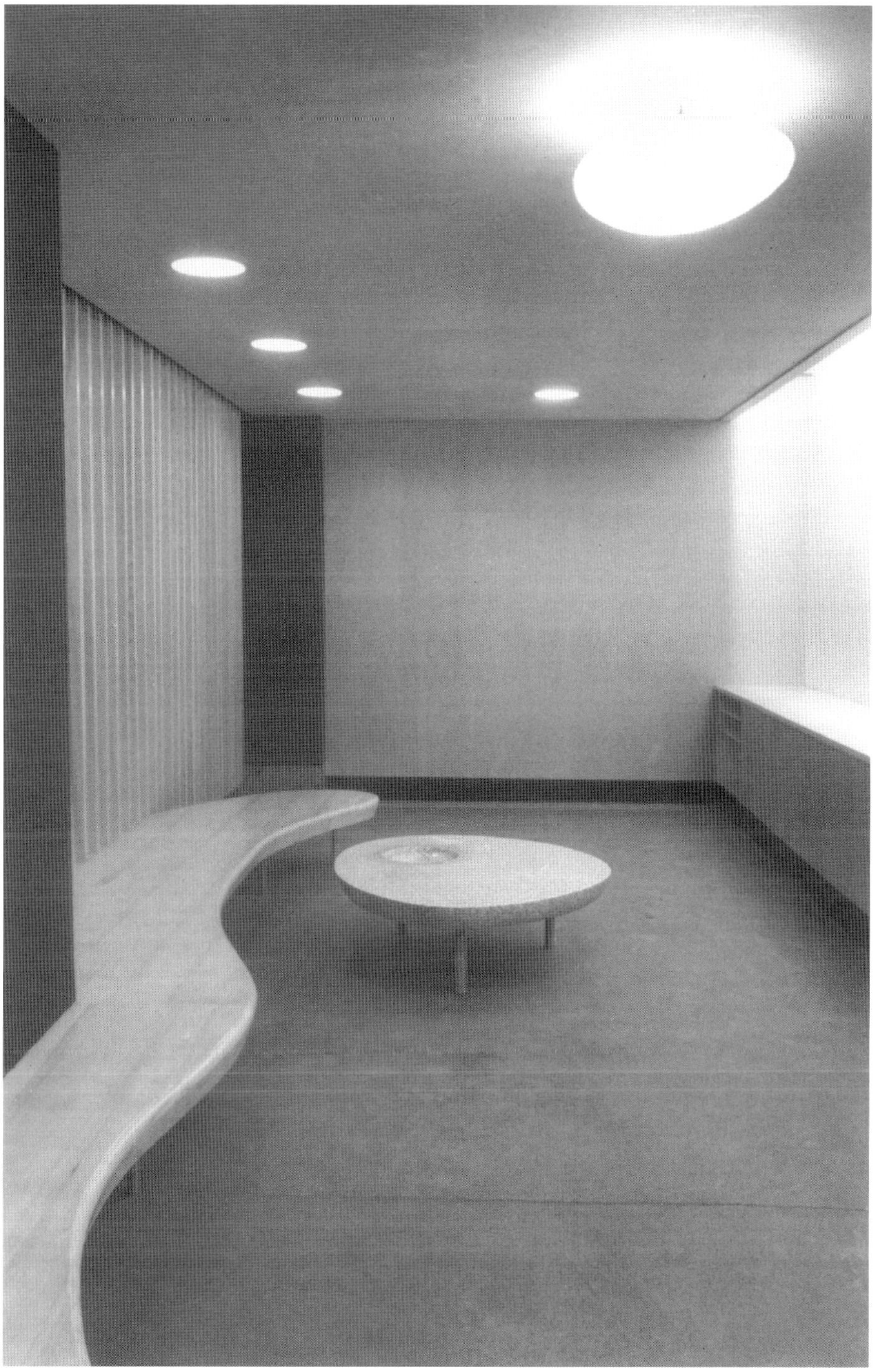

Photographer unknown, Installation View, Chuo Koron Gallery, Tokyo, 1954.

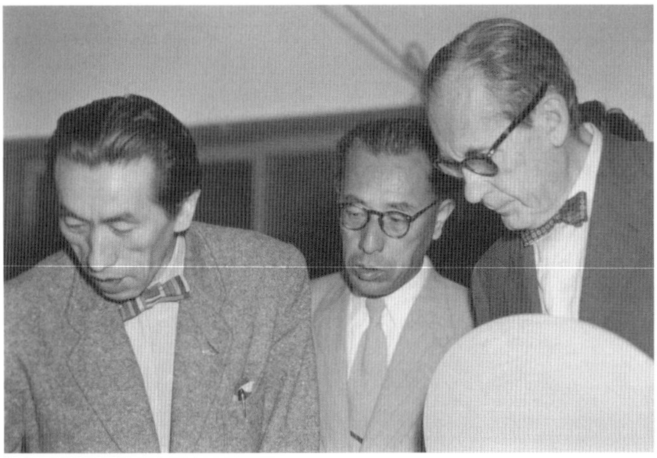

Isamu Kenmochi (left) during a visit from Walter Gropius (right) to the Industrial Arts Research Institute, 1954.

traditional concepts of woodcrafts. The Ministry of Commerce and Industry's objective actively encouraged the growth of plywood and bentwood techniques, the use of new chemicals in bonding adhesives, and standardizing and producing multiple copies of single items.[10]

During the war, in 1940, the IARI was host to another Western pioneer of modernism, Charlotte Perriand (1903–1999), the French designer who had worked in Le Corbusier's studio in Paris during the 1920s. Just as Paris was captured by the Germans, Perriand sailed for Japan at the invitation of two Japanese modernist friends, architects Kunio Maekawa (1905–1986) and Junzo Sakakura (1901–1969), along with an official invitation from the Japanese embassy in Paris asking her to serve as a consultant in industrial design to the Department of Trade Promotion under the auspices of the Ministry of Commerce and Industry. By this time, all national agencies were under the direction of the imperial government.

The three had met as intrepid young architects at Le Corbusier's studio in Paris. Perriand was particularly close to Sakakura.[11] Sakakura returned to Japan in 1936 and shortly thereafter was commissioned to design a Japanese Pavilion for the 1937 Paris World's Fair. This pavilion, which managed to successfully integrate regional characteristics within a modern Le Corbusier–inspired architectural framework, won a Grand Prix at the fair. Perriand wanted to visit her successful colleague and learn more about the Japanese aesthetic and traditional crafts that had fascinated her in her student days.

Although Western nations were arming themselves against the Axis powers, Perriand remained in Japan for several months, meeting with artists and designers, studying the various craft techniques used by farmers, and fiercely debating the Japanese government's export-design policies that effectively conflicted with Japan's aim to colonize all of Asia. Japanese goods after 1937 were not desirable abroad. During her stay in Japan, Perriand encouraged historically significant interchanges among those who had previously been separated by ideological and professional boundaries, generating dialogue and cooperation between the industrial designers and craft-oriented artisans who clustered around the IARI. Perriand keenly supported exchanges between preservationists of indigenous folk arts and export-minded government officials, between architects of national monuments and designers of mass-produced furniture and goods. Kenmochi was impressed with Perriand's apparently intuitive understanding of the problems facing industrial design in Japan. He learned from her appreciation of vernacular Japanese objects, and her detailed assessment of the form, construction, materials, and colors of

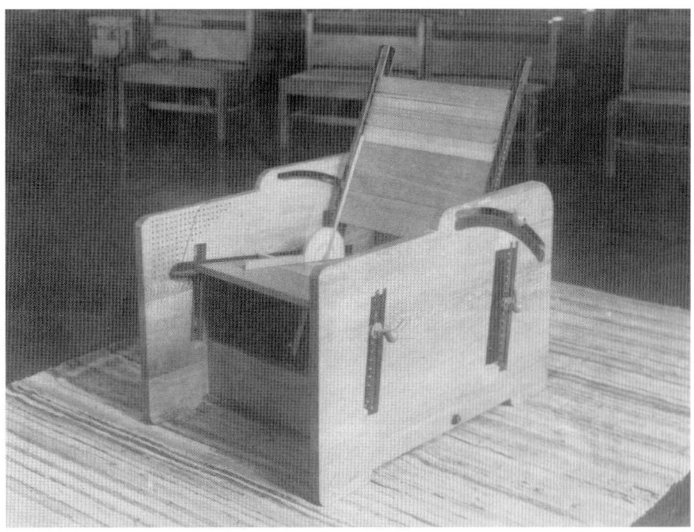

Right: Isamu Kenmochi, chair for standardizing measurements, 1934. Opposite: Yonejiro Noguchi, Isamu's father, 1899.

indigenous Japanese craft traditions. As complicated as these interchanges were for her at the time, her efforts would subsequently alter and redefine intercultural negotiations of Japan's national aesthetics in the postwar period.[12]

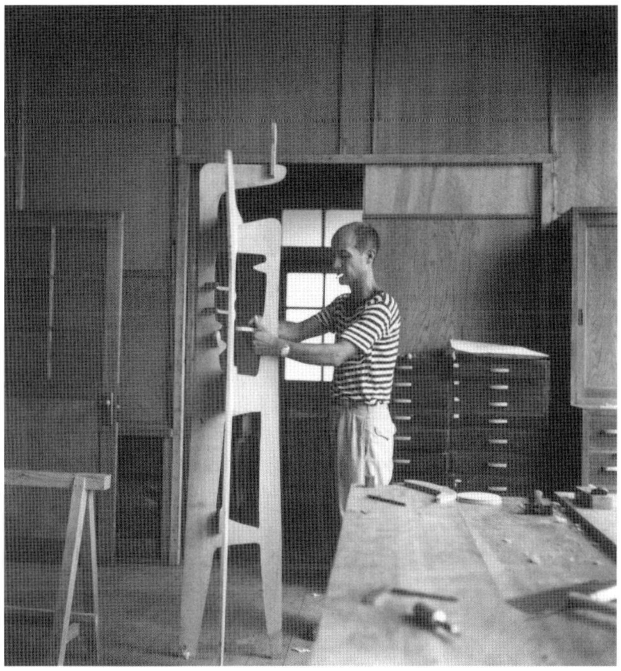

Isamu Noguchi at work on the sculpture *Wakai Hito* at the Industrial Arts Research Institute, August 1950.

Throughout the 1920s and 1930s there was constant tension between the artistic community in Japan and the government. There were continuous antagonisms among the diverse cultural communities as well. Japan continually embraced and then rejected design theories emanating from the West. The Japan International Architectural Association, a nongovernmental group of young architects from Osaka headed by Isaburo Ueno, brashly arranged for Bruno Taut to be one of the first Westerners to visit the Katsura Imperial Villa. The group advocated an agenda of "Japan-ness" (a word that signifies a theoretical-critical analysis of Japanese architecture and culture through a reinterpretation of traditional Japanese architecture in accordance with the principles of modern architecture and the prevailing [Western] International Style). The Japan International Architectural Association opposed the *teikan*, or neo-nationalist, style of architecture prevalent during the 1920s, which it considered merely decorative and inauthentic in its eclecticism.[13] In the context of the growing nationalism of the 1920s, conservative groups and factions followed Frank Lloyd Wright (1867–1959) and his influence on *teikan* theories of architecture. Translated literally as "imperial crown style," it was symbolized by Wright's Imperial Hotel in Tokyo, which he was commissioned to design in 1922.[14] The basic building example of *teikan* featured reinforced concrete with pseudo-historical or appropriated Western Art Deco and Beaux-Arts styles. Even Mexican-styled decorative elements of porous stone, brick or tile detailing, and roofing made in an overstated form of traditional Japanese tiles were commonplace in the *teikan* style.

Under the banner of imperial Japan and the other Axis powers opposed to the International Style, Taut, who worked to stay in the good graces of his host country, tried to reconcile modernism with indigenous Japanese traditions, mediating a complicated struggle for the survival of modernism and the International Style. Apparently one of the reasons Taut's aesthetic ideology was so successful was its sanction of the historical friction between the imperial household aesthetic (*teikan*) and the brusque vulgarism of the shogunate and samurai warrior class that had existed since the sixteenth century.[15] Kenmochi grafted Taut's approach onto something he would subsequently coin as Japanese Modern. Taut showed Kenmochi that Japan should not follow the West in design, but rather that new and original styles should be developed and that the Japanese should create products that would meet and go beyond international standards of excellence.[16]

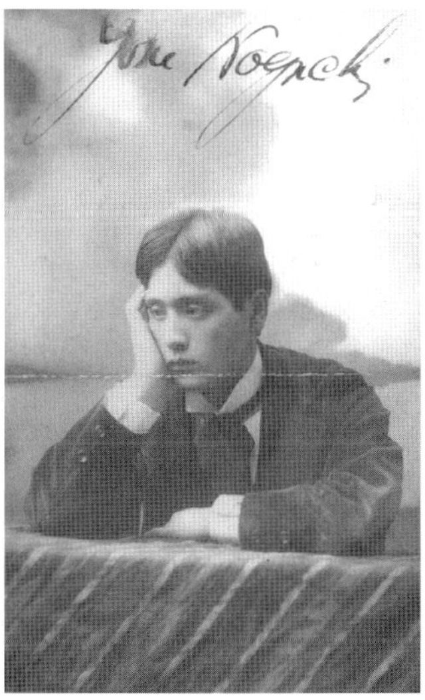

When Noguchi arrived in Japan in May 1950, he was not specifically looking to incorporate the Western design idiom into a postwar-Japan aesthetic. Traveling around the world studying the environments of leisure—a research project funded by the Bollingen Foundation in the United States[17]—Noguchi was intent on finding new strategies to bring art and design into modern life, an ambition he shared

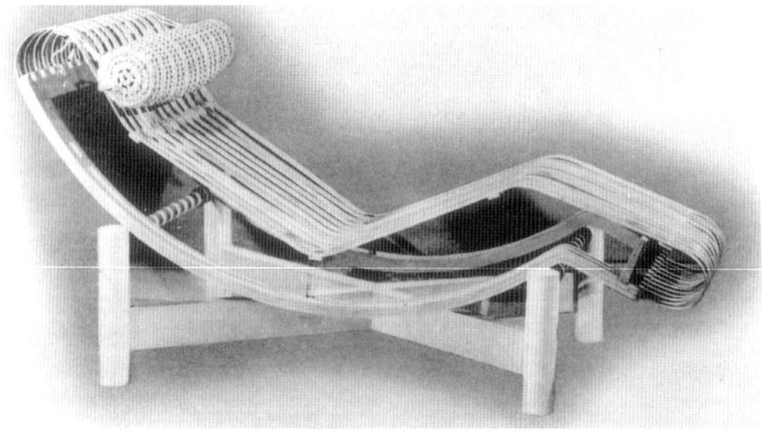

Charlotte Perriand, Bamboo Chaise Longue without seat cover, 1941. Shown at Takashimaya Department Store, 1941.

with Western intellectuals such as Bertrand Russell and Clement Greenberg, who were convinced that while the industrial revolution had increased the number of people who had access to free time, this increase did not apparently produce a popular culture with any sufficient depth or value. Noguchi had only hostility for the materialism of this modern leisure culture, especially as it manifested itself in postwar American excesses. The danger of modern leisure, he believed, was its passivity and tendency towards kitsch.[18]

While the Bauhaus championed the integration of the arts and the diffusion of innovative good design throughout society, Noguchi held the belief that it was the individual artist's responsibility to find a contemporary way into the hearts of the community and that creativity was a product not only of scientific and artistic inquiry but also of a kind of spirituality. Noguchi's design aesthetic was nevertheless fundamentally and firmly grounded in the tenets of Modernism and the International Style, which affirmed the ability of humans to make, improve, or deconstruct and reshape their built and designed environments using new scientific and technological knowledge. Progressive innovation relevant to the cultural context was of ethical importance, as was the notion of newness and originality, a position Noguchi urged his Japanese colleagues to adopt.[19]

On May 6, a few days after his arrival in Japan, Noguchi visited Keio University in Tokyo, where his estranged poet father, Yonejiro Noguchi, had taught for many years.[20] There he met with the distinguished architect Yoshiro Taniguchi (1904–79), with whom he had become acquainted the day before at an art exhibition.[21] The same age as Noguchi, Taniguchi had decided to study architecture after the Great Kanto Earthquake of 1923. The destruction he witnessed from this disaster fortified his desire to design structures that would be safer and stronger than those in existence at the time. In 1928, Taniguchi became a professor of architecture at Tokyo Institute of Technology, developing and implementing hydraulic systems for earthquake-prone buildings. Deeply concerned with architecture facing demolition during postwar reconstruction in Japan, he helped found the Agency for Cultural Affairs, working tirelessly in the restoration and preservation of shrines and temples.[22] Recognized in Japan for his regionalist approach to the International Style, Taniguchi was in charge of reconstructing the university's facilities that had been destroyed during the war.

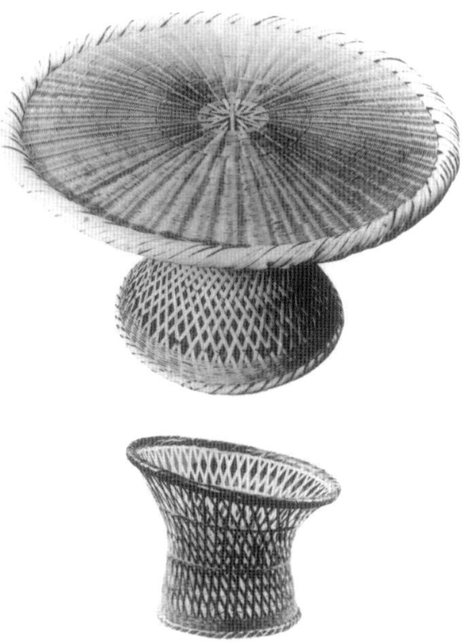

Charlotte Perriand, Table and Wastebasket, woven bamboo, 1941.

That day in May 1950, accompanied by the painter Genichiro Inokuma and the sculptor Kazuo Kikuchi, Taniguchi gave Noguchi a tour of the Keio campus and explained the construction designs for the new campus facilities. They quickly took a liking to one another, whereupon Taniguchi asked Noguchi if he would like to collaborate on the design of the Faculty Hall at Keio. Noguchi eagerly agreed.

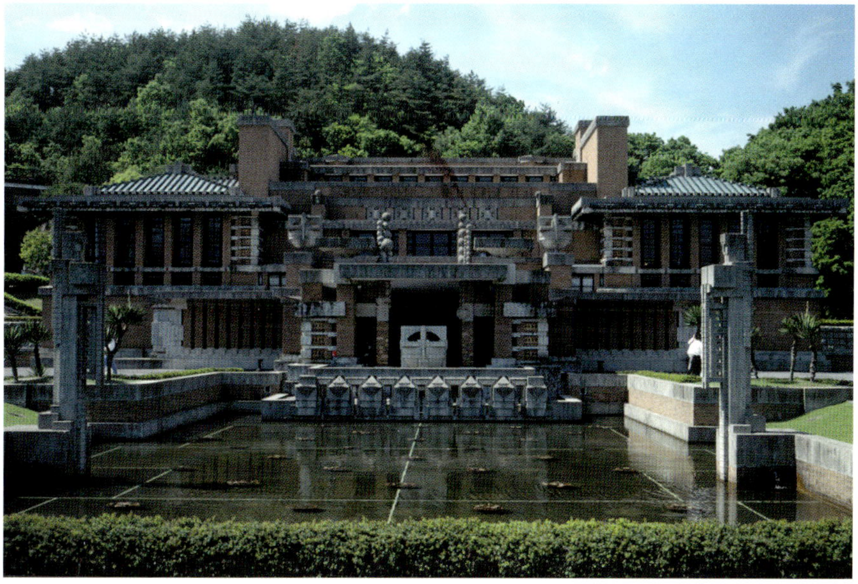

Entrance façade of Frank Lloyd Wright's Imperial Hotel, Tokyo, 1913–1923.

It would take two months for Noguchi to find a studio where he could begin work on the Keio project. In the interim, accompanied by his half-brother, Michio, Noguchi set about finding a release for his creative impulses. Unable to locate a suitable studio for his design project at Keio, he traveled to the Seto region of Aichi Prefecture to work in ceramics, a material he had not handled since his 1930 visit to Japan. Seto had been a ceramics center since ancient times. Hajime Kato, a potter who was later honored with the designation "Living National Treasure," set Noguchi up with supplies, assistants, and a small cottage where he and his brother could live while he worked. In a one-week period, Noguchi produced more than nineteen ceramic works of unglazed Seto stoneware and terracotta.[23]

On June 24, Noguchi was introduced to Kenmochi at the Tokyo University offices of the architect Kenzo Tange (1913–2005). Noguchi had met Tange for the first time shortly after his arrival in Tokyo at a welcoming party organized for him by the newspaper *Mainichi*. Noguchi was well aware of Tange's reputation and professional pedigree, which included winning several architectural competitions—for the Greater East Asia Memorial Building (1942), which was based on the historic Ise Shrine; the Japan Cultural Center in Bangkok (a building praised by Taut), based on the design of the Imperial Palace in Kyoto (1943); and the coveted design for the Hiroshima Peace Park, (1949), in which Noguchi was extremely interested.[24] In fact, Noguchi would subsequently work on several projects in Japan with Tange, including a design for the Peace Park Memorial and bridges connecting Peace Park to Hiroshima City, as well as an interior garden in 1977 for the main headquarters of the Sogetsu Flower Arranging School in the Akasaka area of Tokyo.

Kenzo Tange's design idiom derived from various influences, among them Frank Lloyd Wright and Hideto Kishida (1899–1965), his professor at Tokyo Imperial University. Many of Tange's designs used Japanese works of historical significance as models. He learned from the International Style yet incorporated conventional Japanese proportions based on such iconic and traditionally crafted structures as the Katsura Imperial Palace in Kyoto and the Ise Shrine. Kishida promoted Japanese design by means of a modernist intervention, capturing elements of past architectural artifacts and reinventing them for a present context. This austere and reductionist approach taught Tange that certain Japanese aesthetics were compatible with modern design.[25]

At Tange's studio, Kenmochi suggested that Noguchi visit him at the Industrial Arts Research Institute in Tokyo, where Kenmochi was serving as a technical officer, and look into its facilities. Desperately in need of a studio, Noguchi asked if he could work at the Institute, and Kenmochi obliged.

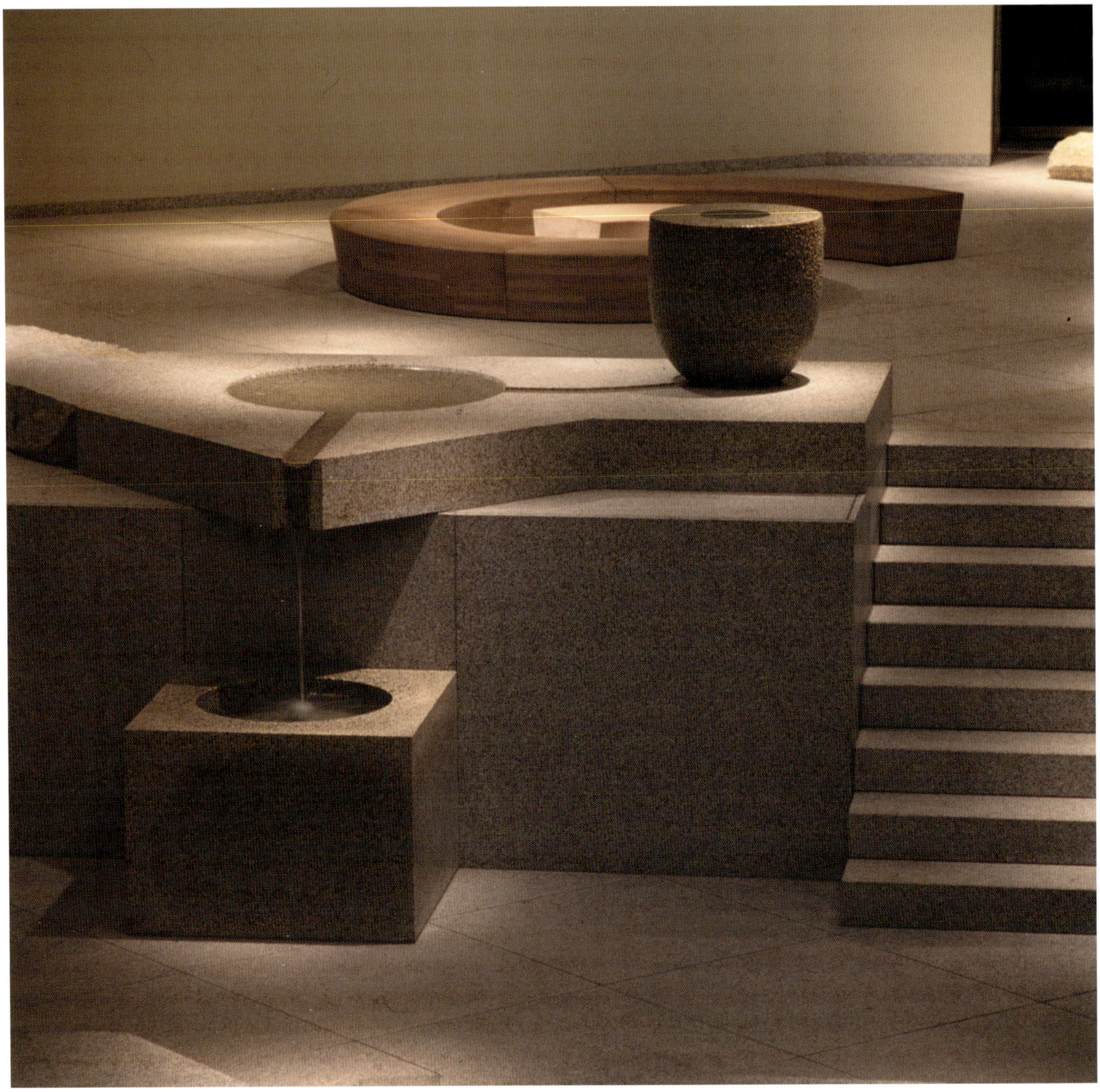

Isamu Noguchi, *Tengoku*, Interior Garden, Sogetsu Flower Arranging School, Tokyo, 1977–1978.

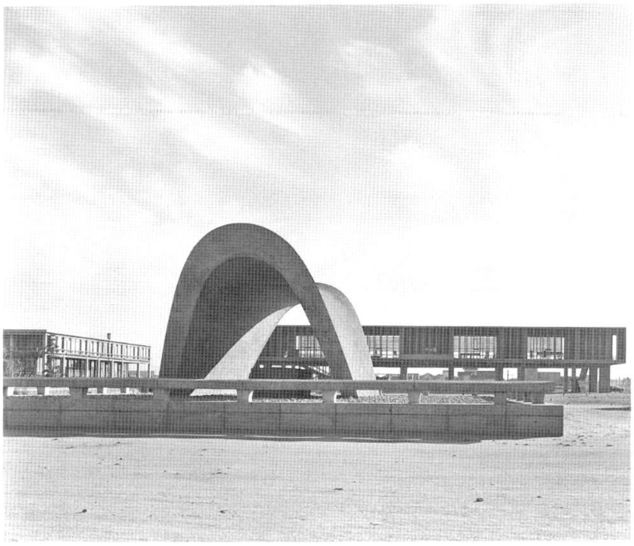

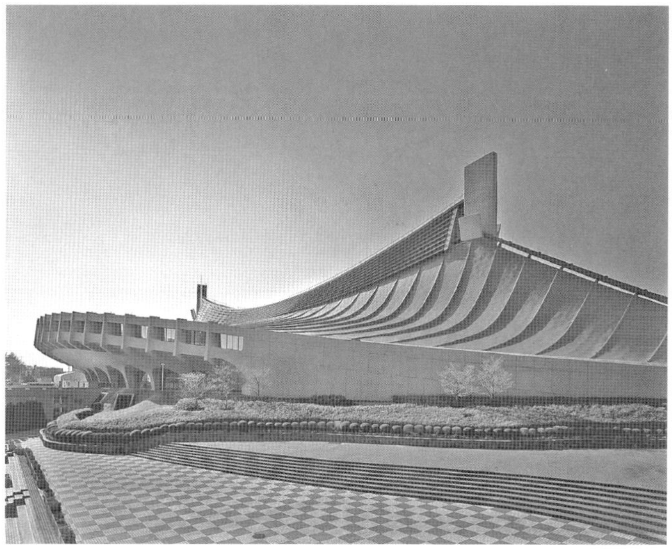

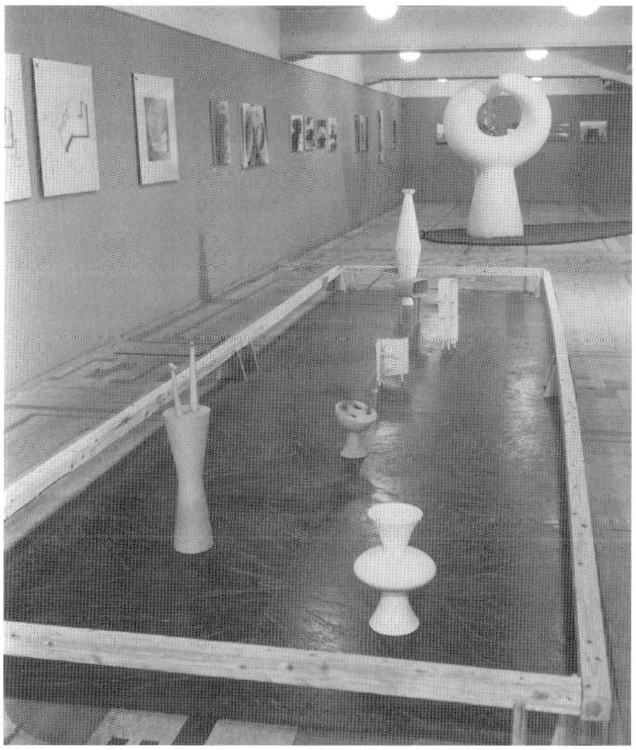

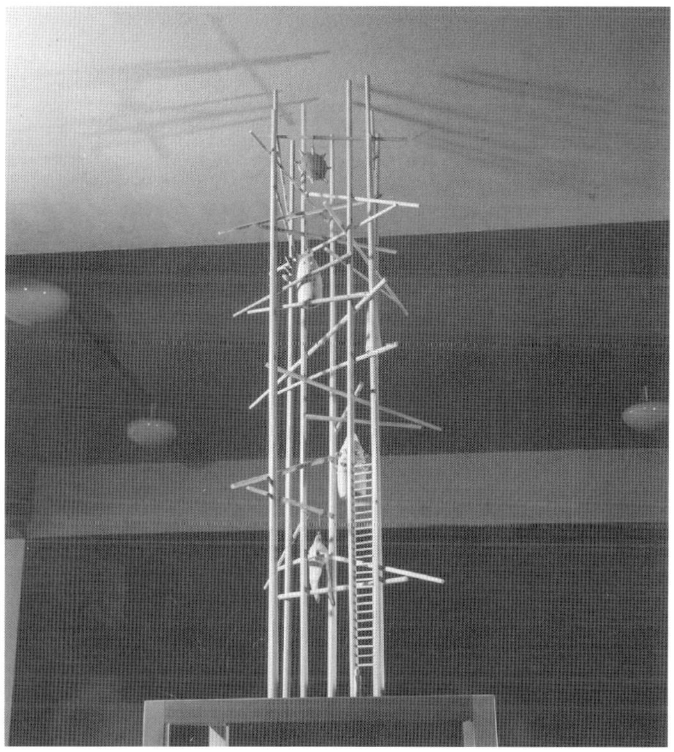

Clockwise from top left: Kenzo Tange, Hiroshima Peace Memorial Museum and Cenotaph, Hiroshima, Japan, 1949–1955; Kenzo Tange, National Olympic Stadium, Yoyogi, Japan, 1961–1964; Isamu Noguchi, *Model for Bell Tower for Hiroshima*, 1950; exhibition of Noguchi ceramics and sculpture at Mitsukoshi Department Store, 1950.

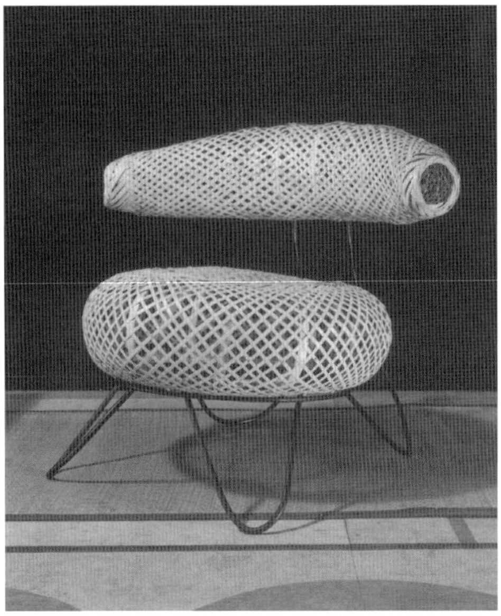

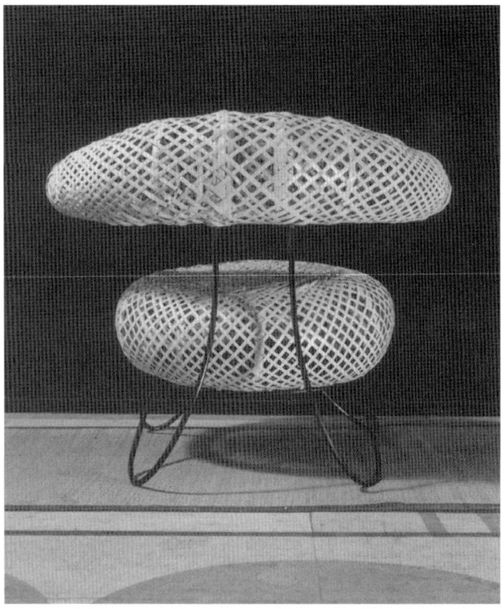

Bamboo Basket Chair prototype designed by Isamu Noguchi and Isamu Kenmochi, 1950.

Noguchi was in a good position to observe the irony that the Japanese were neglecting the rich treasures of their own heritage to imitate American models, often resulting in unsatisfactory products, whereas a prevailing wave of Japanism in the United States led some of the best American designers and artists to produce superior modern work on the basis of their research in Japanese tradition.[26] While working at the IARI, Noguchi took an idiosyncratic interest in the experimental furniture designs Kenmochi was creating.

One notion Kenmochi explored was that of using the structure and techniques of Japanese bamboo basket weaving as the basis of a design for the seat of a chair. Noguchi was intrigued by this chair, and offered to collaborate with Kenmochi. Noguchi immediately drafted plans, and Kenmochi engaged an expert bamboo craftsman to weave strips of bamboo into the shapes Noguchi had devised. Kenmochi originally envisioned the frame for this chair to be of wood but Noguchi's idea was to use a single continuous bent iron rod that would form the legs as well as the full supporting armature for the bamboo elements. Unlike the one-of-a-kind furniture pieces Noguchi was making for the Keio University project, the Bamboo Basket Chair was designed as a prototype for mass-production.[27] It is speculated that several models of the bamboo chair were made but ultimately the piece did not go into production, probably because it was deemed not sufficiently durable.[28] Nevertheless, the process of creating the experimental bamboo chair was dedicated to the purpose of advancing the modernization of Japanese forms in tandem with the use of traditional Japanese techniques and materials which could be brought together for a Japanese and foreign market.

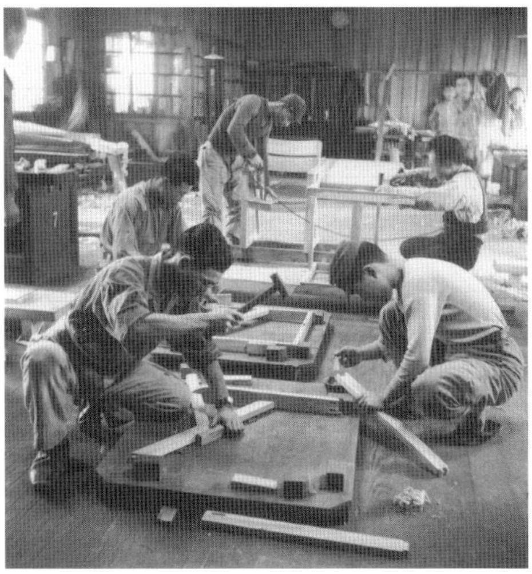

Staff at the Industrial Arts Research Institute in Sendai making furniture for the U.S. Army officers' club, 1948.

During a brief two-week period in August 1950, Noguchi commuted daily to the Institute where he furiously worked on his plans for the Faculty Room for Keio University. In this short period of time, he constructed the basket chair, benches, stools, a table, and two sculptures, as well as a wall relief. He also worked on a project he initiated for a bell tower for Tange's Hiroshima Peace Park, constructing two models out of bamboo and ceramic at IARI.

Earlier that summer, on June 19 and again on July 10, Noguchi gave a lecture titled "Art and the People" to a meeting of the Japanese Artists' Federation, a loosely formed union of painters and sculptors who agreed to endorse Noguchi's design of *Bell Tower for Hiroshima*. It is not known if Tange fully supported Noguchi's formulation of these models, or if it was solely an ambitious strategy on the part of Noguchi in anticipation of a commission for Hiroshima.[29] These models, along with the furniture and sculpture for Keio and a group of ceramics Noguchi had made earlier in the summer, would be presented in an exhibition at the Mitsukoshi Department Store in Tokyo on August 18–27, 1950.

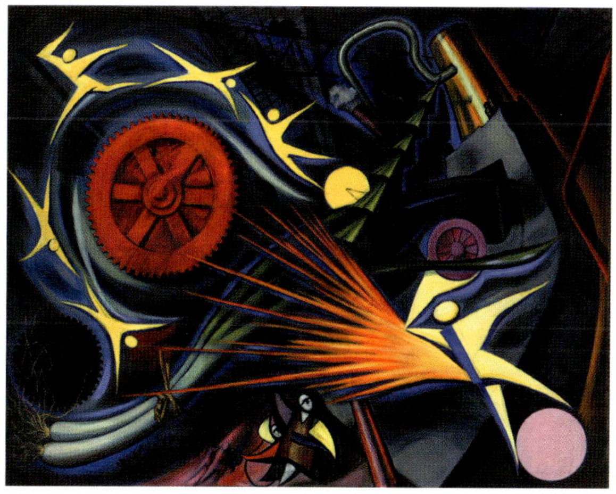

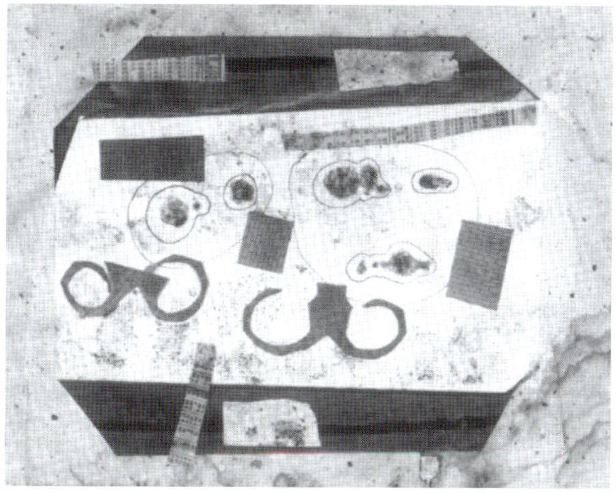

Top: Taro Okamoto, *Heavy Industry*, oil on canvas, 1949.
Above: Saburo Hasegawa, *Newspaper Collage*, 1937.

Noguchi was not accustomed to the tradition of holding serious exhibitions at department stores, and he initially resisted the idea. The Mitsukoshi Department Store in the Nihonbashi neighborhood was one of the largest and most unusual department stores in Tokyo, and had a long history of holding important exhibitions. Taniguchi designed the solo exhibition for Noguchi, sparing no effort to give Noguchi's objects sensitive surroundings. Taniguchi modified the floor and surfaces in each gallery in order to obtain the optimum aesthetic conditions for displaying the works. In an era when even exhibitions of Japanese paintings imitated the West, the installation of Noguchi's sculpture at Mitsukoshi in 1950 conveyed a vague Japanese sensibility, employing minimal structures and some works placed directly on the floor. The ambiance of the exhibition design offered a sense of calm and repose, while areas such as those surrounding the plaster of *Mu*, a sculpture intended for the garden at Keio, delineated a subsequent landscape shape on the gallery floor. The feeling was of something Japanese-like, though the sensibility was not traditionally Japanese.

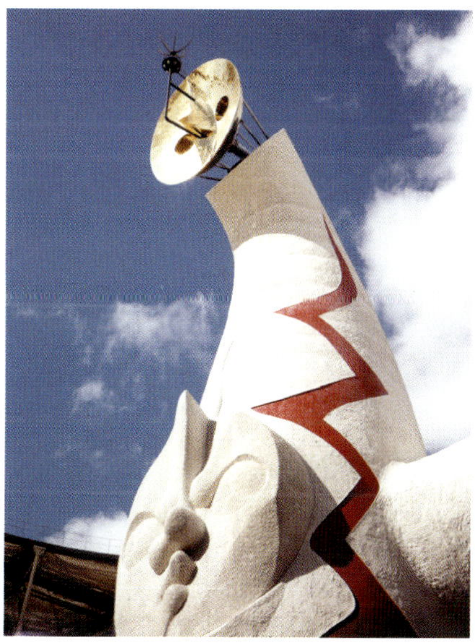

Taro Okamoto, *Tower of the Sun*, monument for Expo '70 in Osaka.

Although Charlotte Perriand had an exhibition of her furniture designs at the Takashimaya Department Store in Tokyo in 1941, it would be five years after the war ended before the Japanese art community would see important exhibitions by artists from abroad. Besides Noguchi in 1950, there would also be one Matisse and two Picasso exhibitions at the Mitsukoshi store in 1951.[30]

When Noguchi arrived in Japan, after the war, he spoke out against the subjugation of art to nationality, placing himself unequivocally on the opposite side of the political spectrum from his militaristic father, who had died of stomach cancer. Yonejiro Noguchi had begun his vociferous support of Japanese militant imperialism and its colonial ambitions early in the 1930s, which would alter his international standing as a learned gentleman and poet. Father and son never entirely reconciled their estrangement.

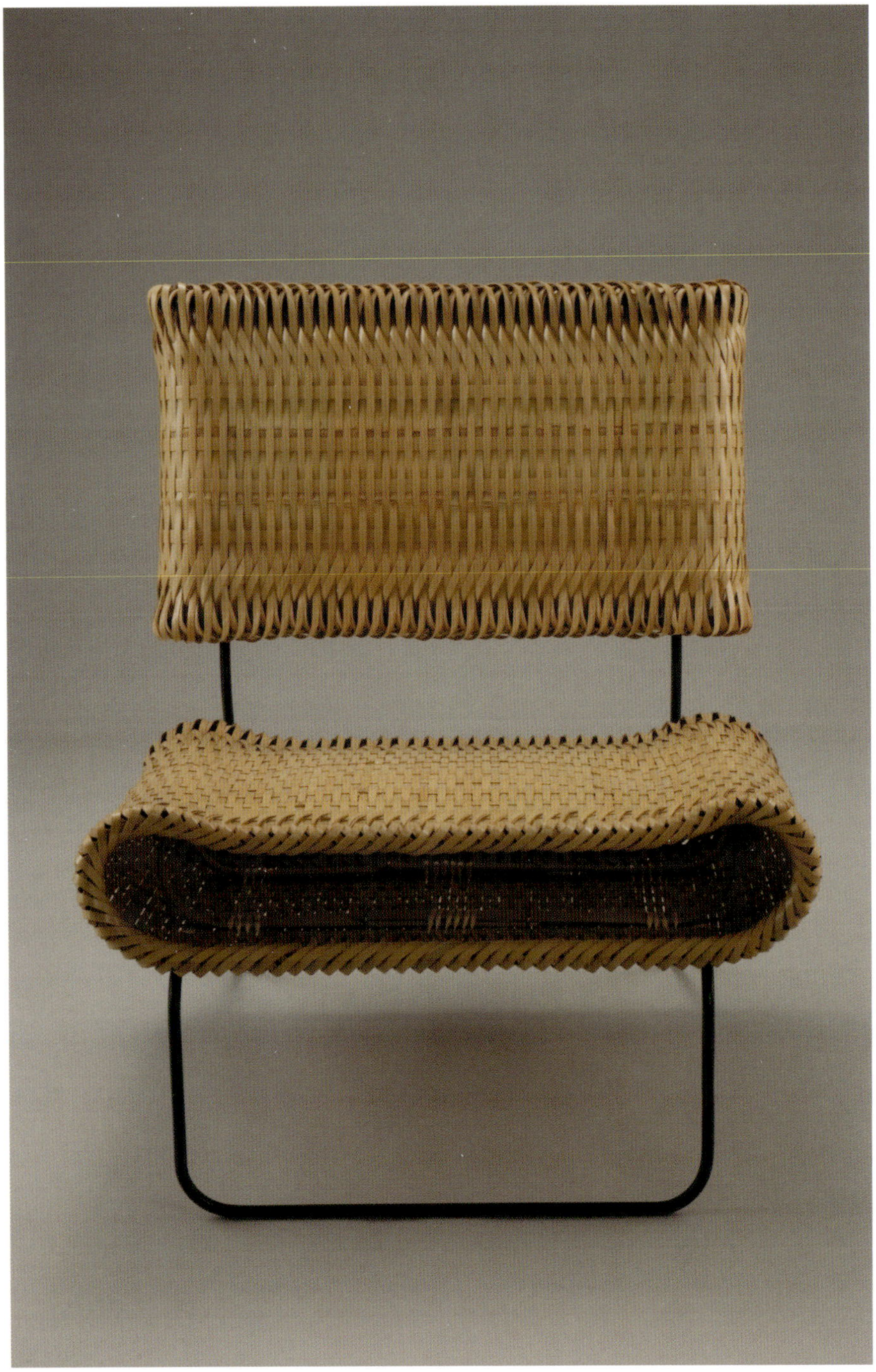

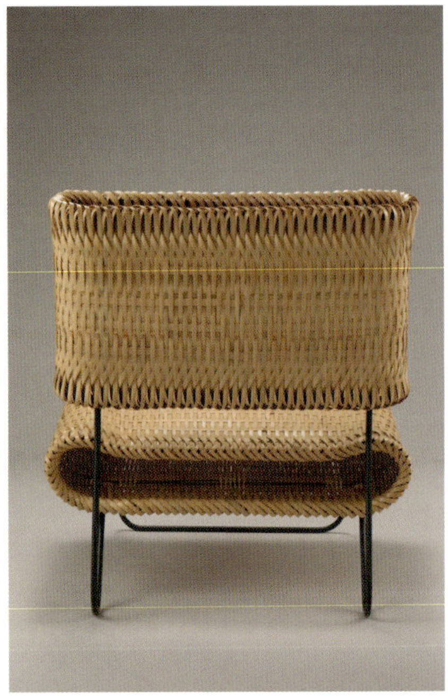

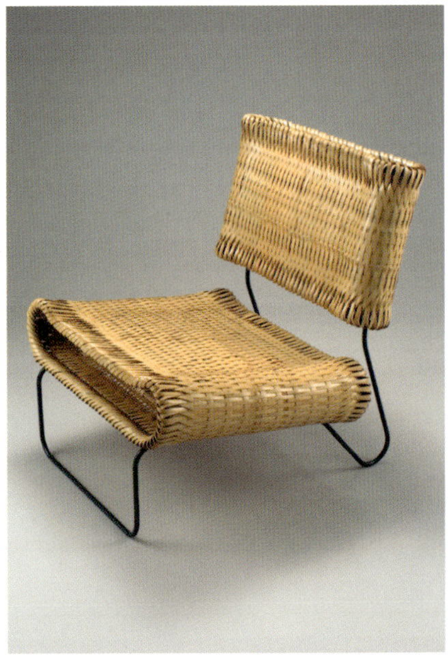

Isamu Kenmochi, Bamboo Chair, 1950. Exhibition replica 2007 by Kenmochi Design Associates in association with Y.M.K. Co., Ltd., bamboo and iron, 29 9/16 x 27 1/2 x 30 7/16 in.

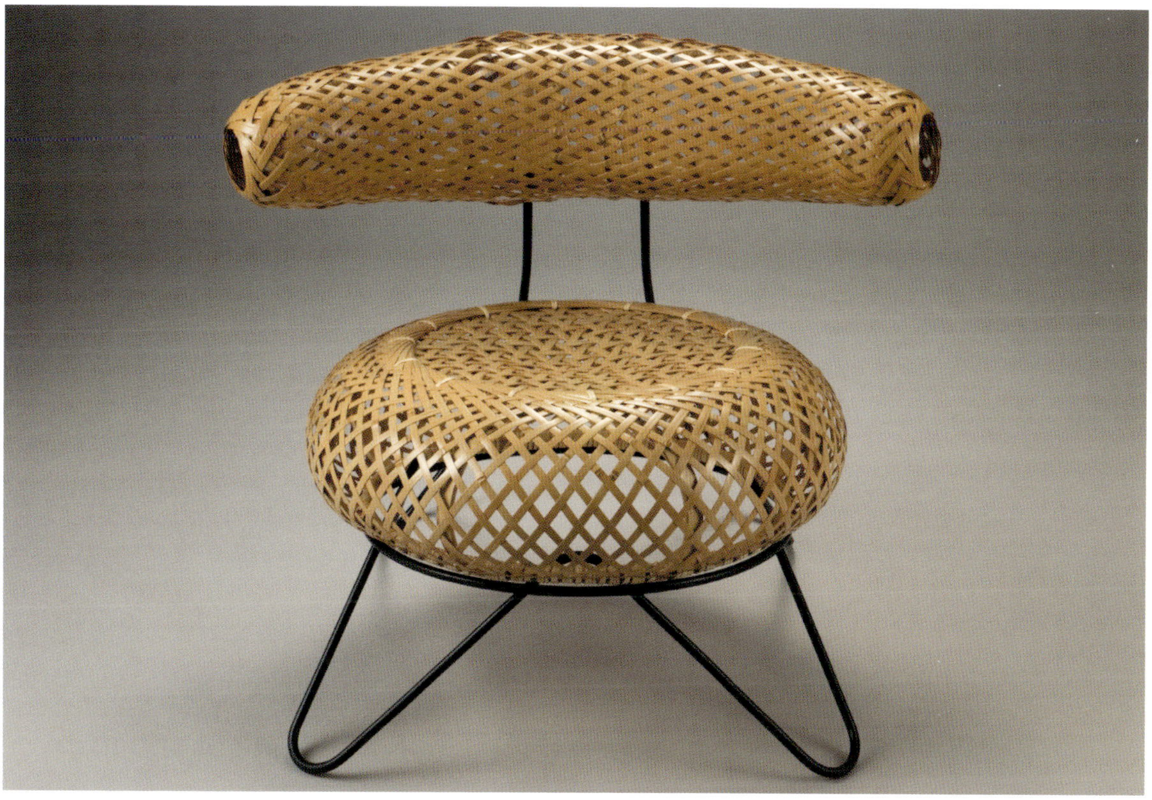

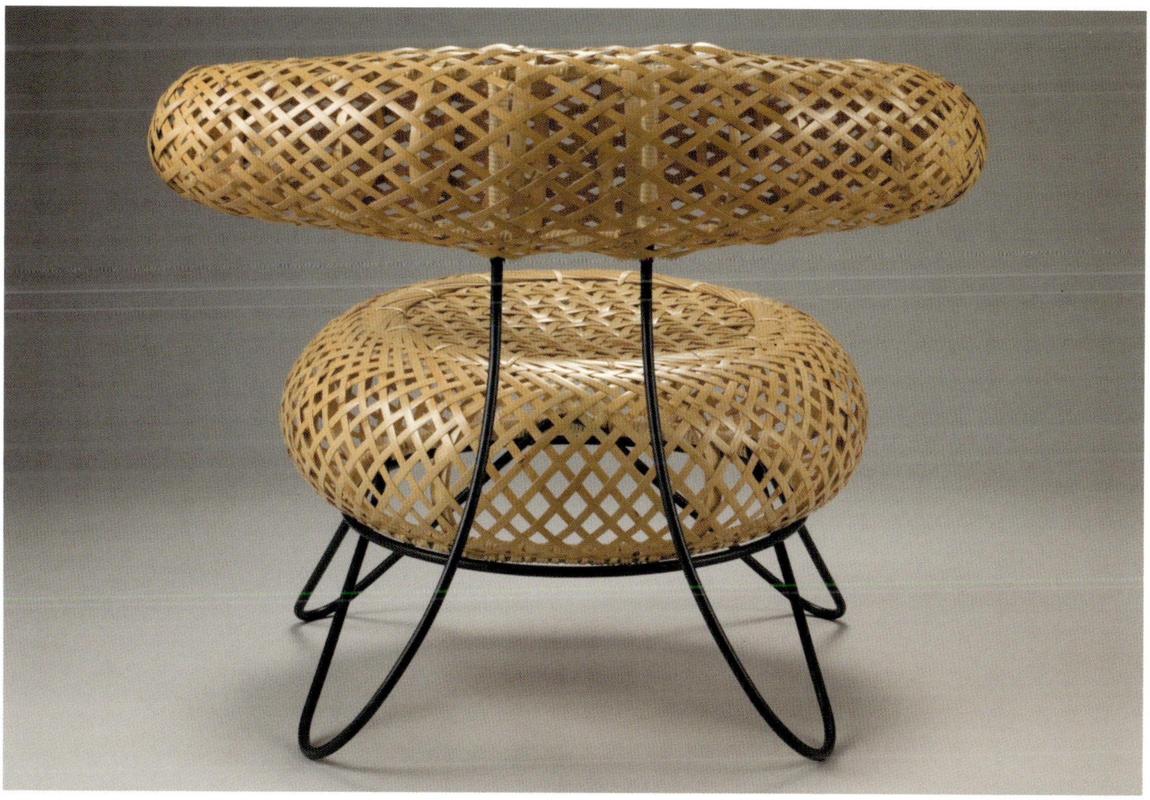

Isamu Kenmochi and Isamu Noguchi, Bamboo Basket Chair, 1950. Exhibition replica 2007 by Kenmochi Design Associates in association with Shosaku Kondo and Y.M.K. Co., Ltd., Shioji, bamboo, and iron, 27 3/8 x 34 1/4 x 29 3/4 in.

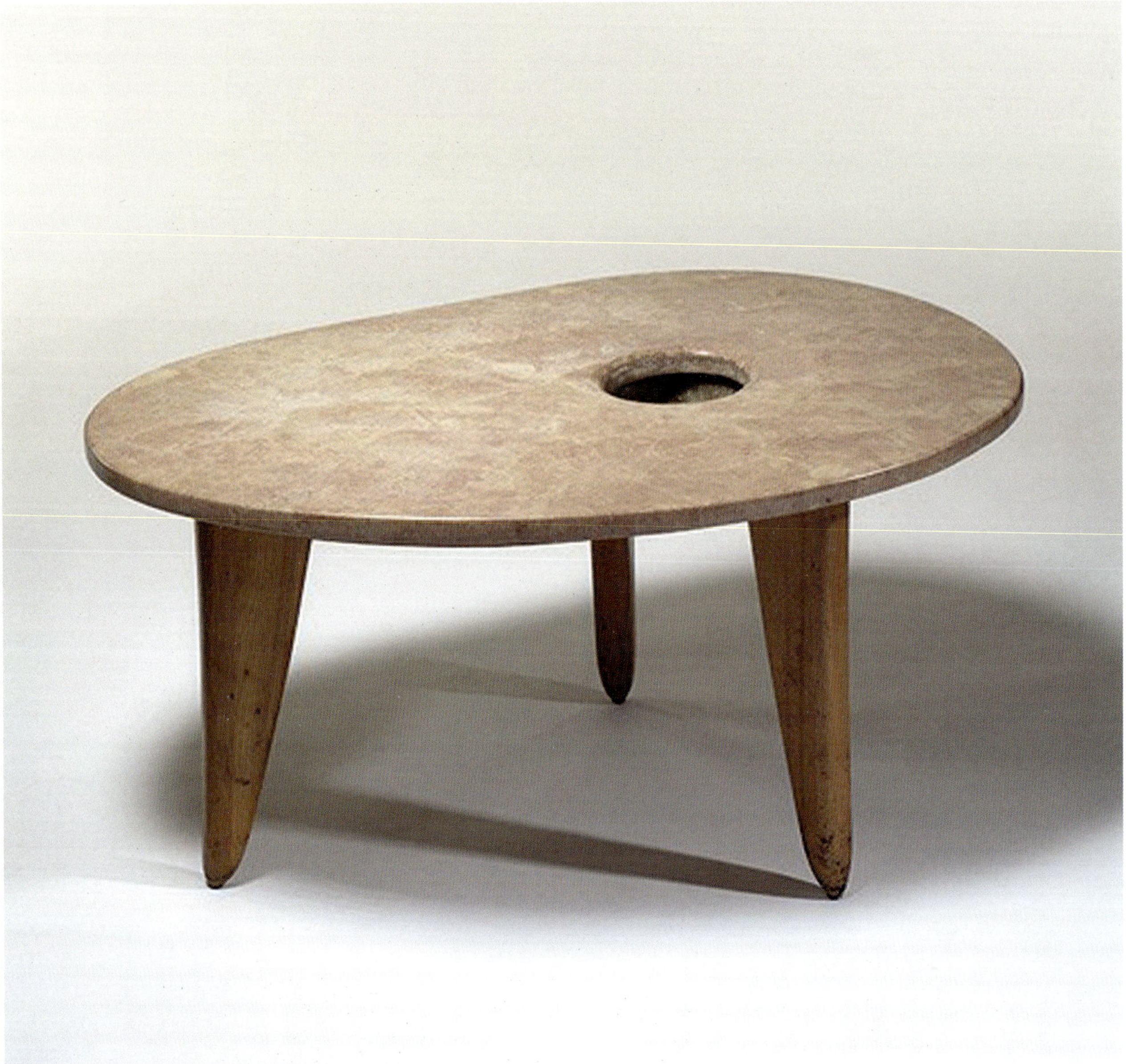

Isamu Noguchi, Marble-Topped Coffee Table, pink marble, maple, plywood, and aluminum, 1948–1950, 18 3/4 x 38 1/4 x 26 1/2 in. The Marble-Topped Coffee Table is one of the most personal and artistic of the Noguchi furniture designs marketed by the Herman Miller Co. Only a few examples of the table were produced.

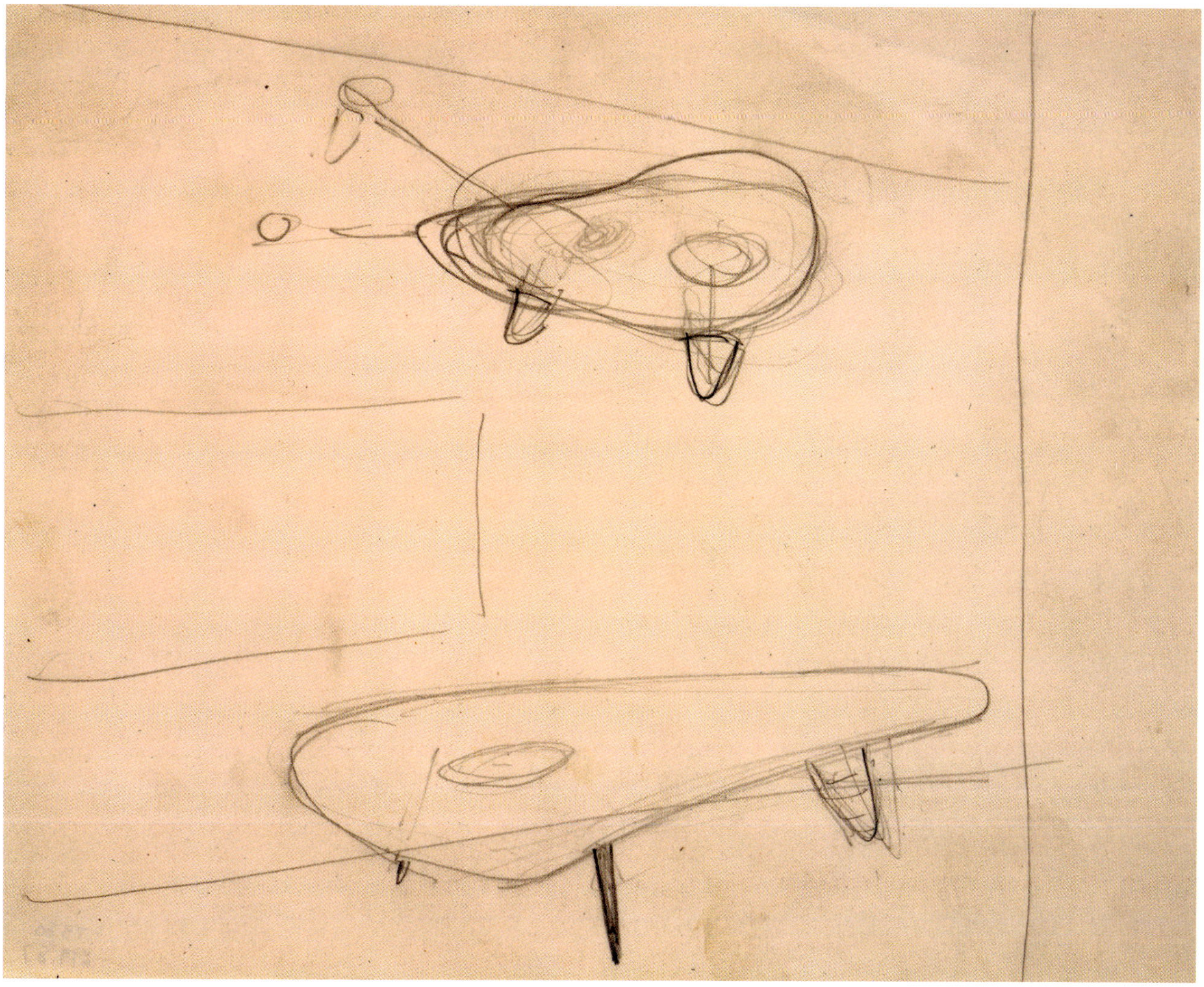

Isamu Noguchi, Table and Lamp, pencil on paper, ca. 1944, 8 1/2 x 11 in.

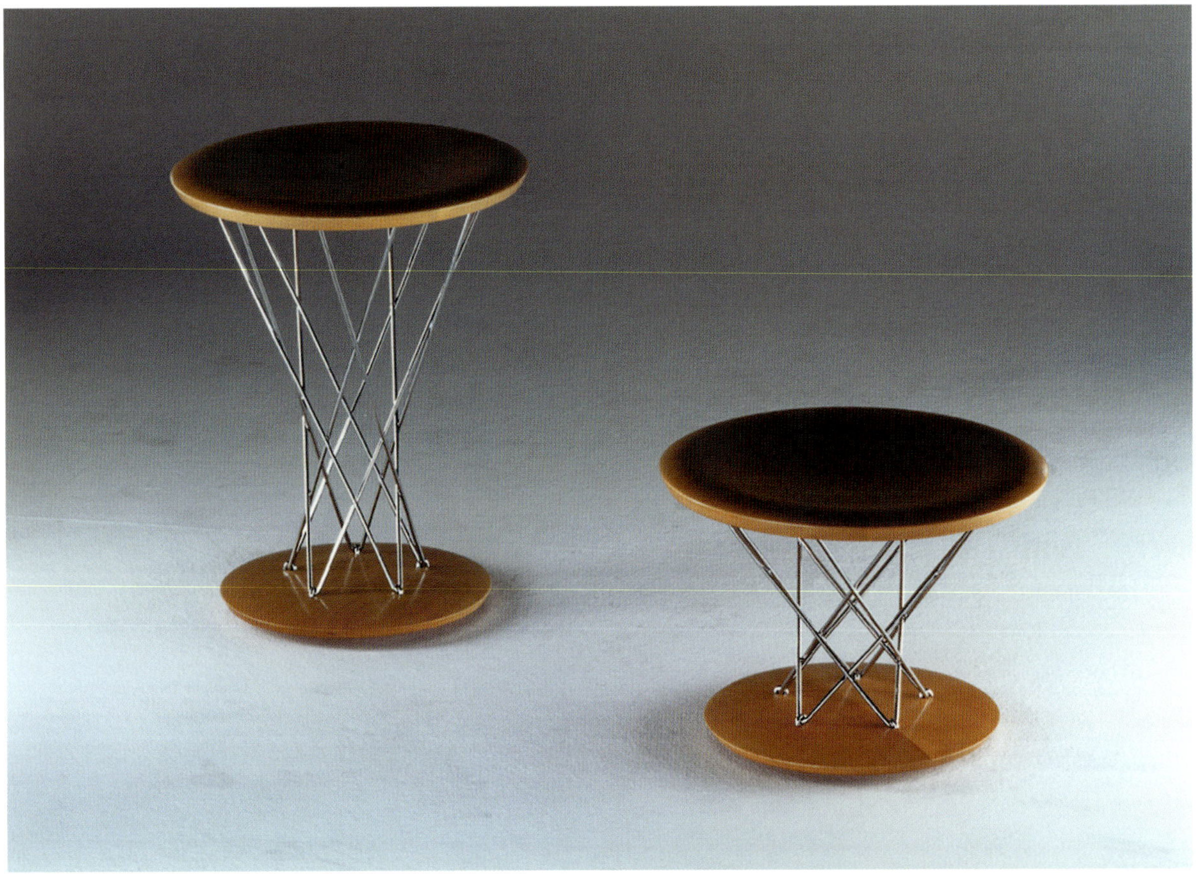

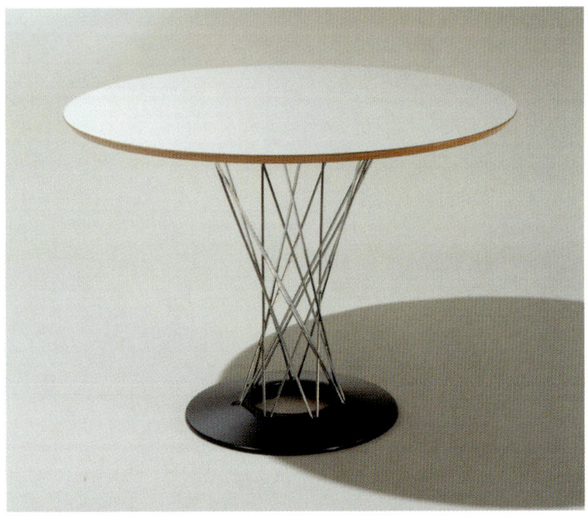

Top: Isamu Noguchi, Rocking Stools, chrome-plated steel wire and birch, 1954, small: 8 1/4 x 13 in., large: 16 3/4 x 14 in. The Rocking Stool and the Wire Table, manufactured by Knoll Associates, Inc., had their roots in the postwar vogue for wire furniture. The use of iron rods and steel struts exploited both the strength and decorative possibilities of the materials. Bottom: Isamu Noguchi, Large Dining Table, plastic laminate, Finland birch plywood, chrome-plated steel wire, and cast iron with black porcelain finish, 1955, 28 1/2 x 48 in.

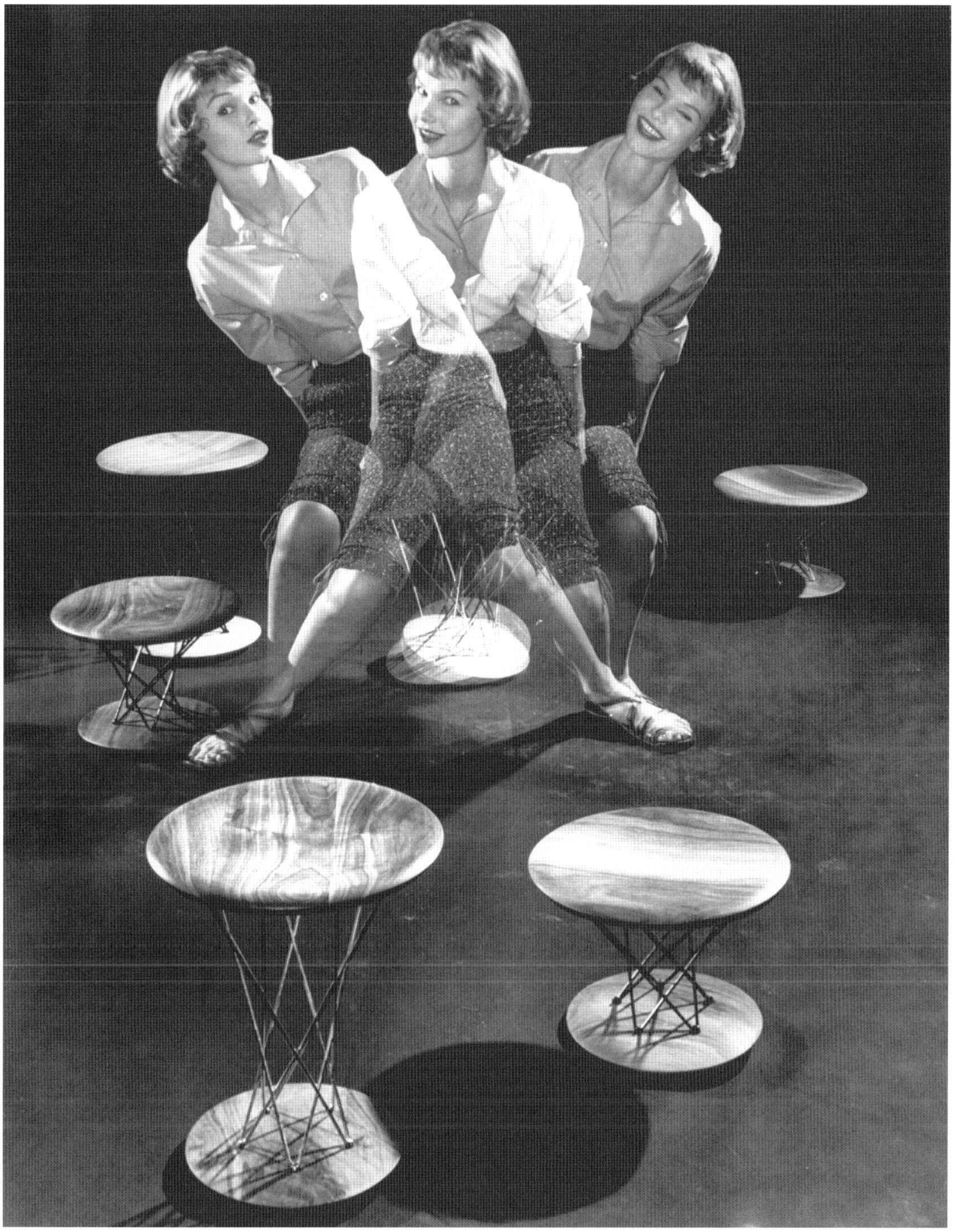

Unknown photographer, Knoll Associates publicity photograph, A Model on Noguchi's Rocking Stool, 1956.

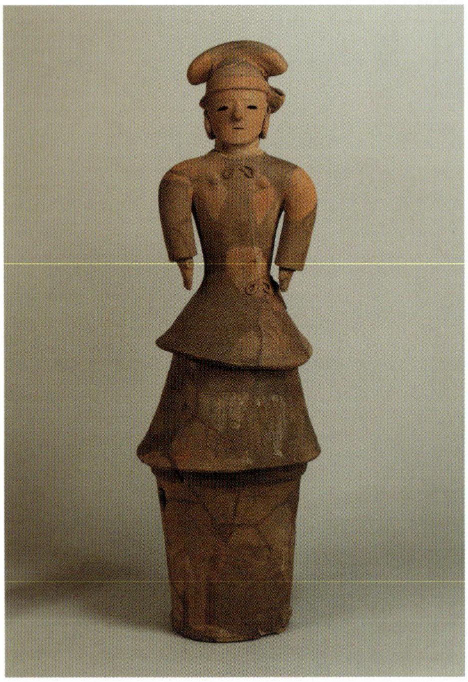

Haniwa sculpture known as *Woman in Full Dress*, earthenware, sixth century.

Along with his appreciation of Japanese traditions, Noguchi voiced his opposition to cultural nationalism, and his public opinions expressed in lectures and interviews can be read as a critique of his father's notorious nationalistic excesses. On the other hand, there is no question that being the son of a famous poet was not lost on the ambitious younger Noguchi. In a country where Isamu Noguchi did not speak the language—despite his having lived in Japan with his Caucasian mother until he was thirteen years old, before moving to the United States—the association with his father opened many professional doors for him.

On May 11, 1950, soon after his arrival in Tokyo, Noguchi met with the Japan Avant-Garde Artists Club, which had been formed in 1947 by a group of artists and critics associated with Surrealism and abstraction. Among the members were Taro Okamoto and Saburo Hasegawa, who joined a lecture sponsored by the newspaper *Mainichi* at the Mainichi Hall in Yurakucho, Tokyo. Noguchi's lecture was titled "Art and Community," and Hasegawa served as the moderator. This club, a Communist organization that rejected Modernism in the name of Socialist Realism, existed in opposition to the Japan Art Association.[31]

Having followed the work of Isamu Noguchi for several years before they actually met, Saburo Hasegawa (1906–1957) assumed the role offered him by the *Mainichi* newspaper to escort Noguchi through Japan. Ultimately he became spokesman, translator, and guide for Noguchi during his stay. Noguchi would describe Hasegawa, who was both a painter and a writer, "as painting more or less like Paul Klee." At this same time, Noguchi decided he wanted to travel around Japan and learn more about the country's history and temple traditions, particularly Zen gardens, so he and Hasegawa took a two-week trip that resulted in long discussions of Zen poets, literature, architecture, calligraphy, and tea.[32]

In the 1920s and 1930s, as a result of the influence of the painter and writer Taro Okamoto (1911–1996), Jomon-era artifacts were rediscovered and popularized in Japan among the cultural elites. Noguchi's 1930 visit to Japan afforded him the opportunity to experience the resurgent Japanese interest in the country's ancient pots and utensils, as well as the sophisticated terracotta *haniwa* figurines of the Jomon era (10,000 to 300 B.C.). Noguchi approached Jomon artifacts from a Western Modernist viewpoint, formulating his first artworks inspired by and pregnant with this archaic earthy Japanese style or perhaps directly appropriating this ancient art form.[33]

Okamoto's philosophy stressed the dynamic beauty of the native and rustic Jomon aesthetic. Similar to Noguchi, who felt an urge to learn more about the role of art in older societies, Okamoto turned his attention to the regeneration of Japanese culture after the war. As with Kenmochi, but perhaps not in the case of Noguchi or Kenzo Tange, Okamoto rejected all inclinations to appropriate primitive or ancient forms into any kind of contemporary art.[34] Having also lived in Paris, Okamoto was an incisive

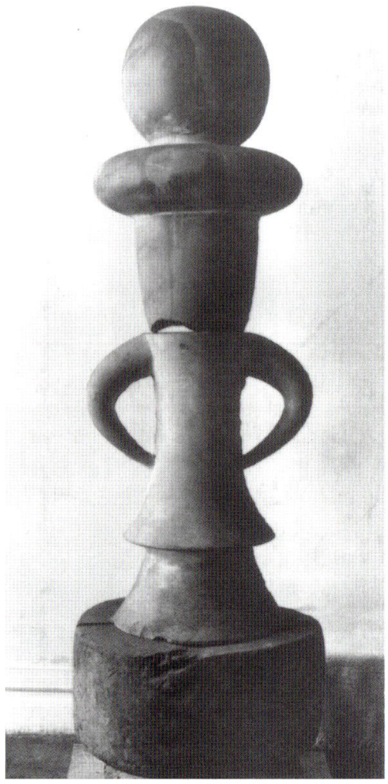

Isamu Noguchi, *The Queen*, terracotta and plaster, 1931.

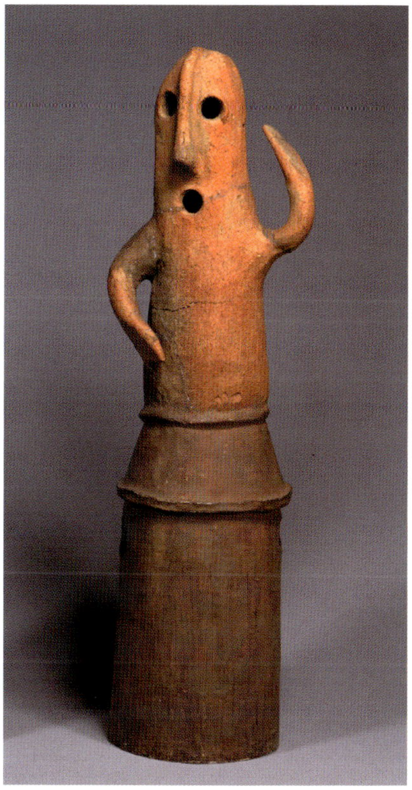

Haniwa sculpture known as *Dancing Man*, earthenware, seventh century.

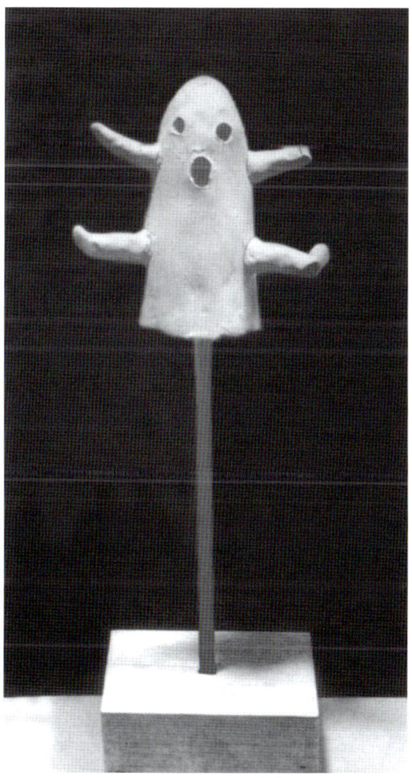

Isamu Noguchi, *Bell Child*, unglazed Seto white stoneware, 1950.

critic of the Western view of Japan and its dangers for postwar Japanese culture. His influence on customs extended to the interactive roles of tradition and contemporary Japanese culture in the development of art and design, a concept that Kenmochi understood. Like Noguchi and other modern artists who studied the "primitive" in other cultures, Okamoto believed that modern art was a universal property, and his model of the tradition of Jomon art was something to be creatively reinvigorated through a dynamic philosophical dialogue. The Jomon-era revival was an extremely important subject of discourse for Japanese artists struggling to find their own identity.[35] The artists' distinctive styles can be seen in their production for Expo '70 in Osaka, where Noguchi, Kenmochi, and Okamoto all presented their work. Kenmochi offered up a public-use bench design that is still in production today. Okamoto created the still-standing *Tower of the Sun*, and Noguchi designed nine fountains, which no longer operate.

Not surprisingly, Noguchi's largely foreign understanding of Japan was formed by the political and cultural events that raged after the Meiji restoration, with its imperial aspirations, and by the array of aesthetic theories that followed. But those influences naturally molded the perceptions of Kenmochi, a native son. Beginning in the mid-nineteenth century, Japan was constantly responding and reacting to the gaze of the West. European and American desires for exotic items from the East resulted in collector's terms like "chinoiserie" and "japonaiserie." Since there were no standards in Japan by which these collectibles could be determined to be good or authentic, a kind of connoisseurship developed in Japan to satisfy this Western desire. Before japonaiserie was a concept, functional objects such as pottery, furniture, and utensils were not considered artworks at all, and craftsmen were not considered artists. Japanese connoisseurs were created by the need to distinguish what was art from ordinary functional objects based on Western tastes.[36]

After his 1950 visit to Japan, Noguchi would return to the country often. The following year would prove intensive, with Noguchi working in Japan on a garden for the Reader's Digest Building in Tokyo—with the architect Antonin Raymond, who had assisted Frank Lloyd Wright on the design of the Imperial Hotel in 1922. Noguchi would also complete the Faculty Room at Keio University with Yoshiro Taniguchi, and continue to press Kenzo Tange and the Hiroshima City officials on his plans for a memorial in Peace Park. In spring 1951, when he returned to Japan, he also created the first designs for his Akari lanterns, the mulberry paper and bamboo lamps that are ubiquitous today and still manufactured in the cottage-style industries of Gifu, Japan. Kenmochi would go on to open his own offices in 1955. Kenmochi Design Associates in Tokyo

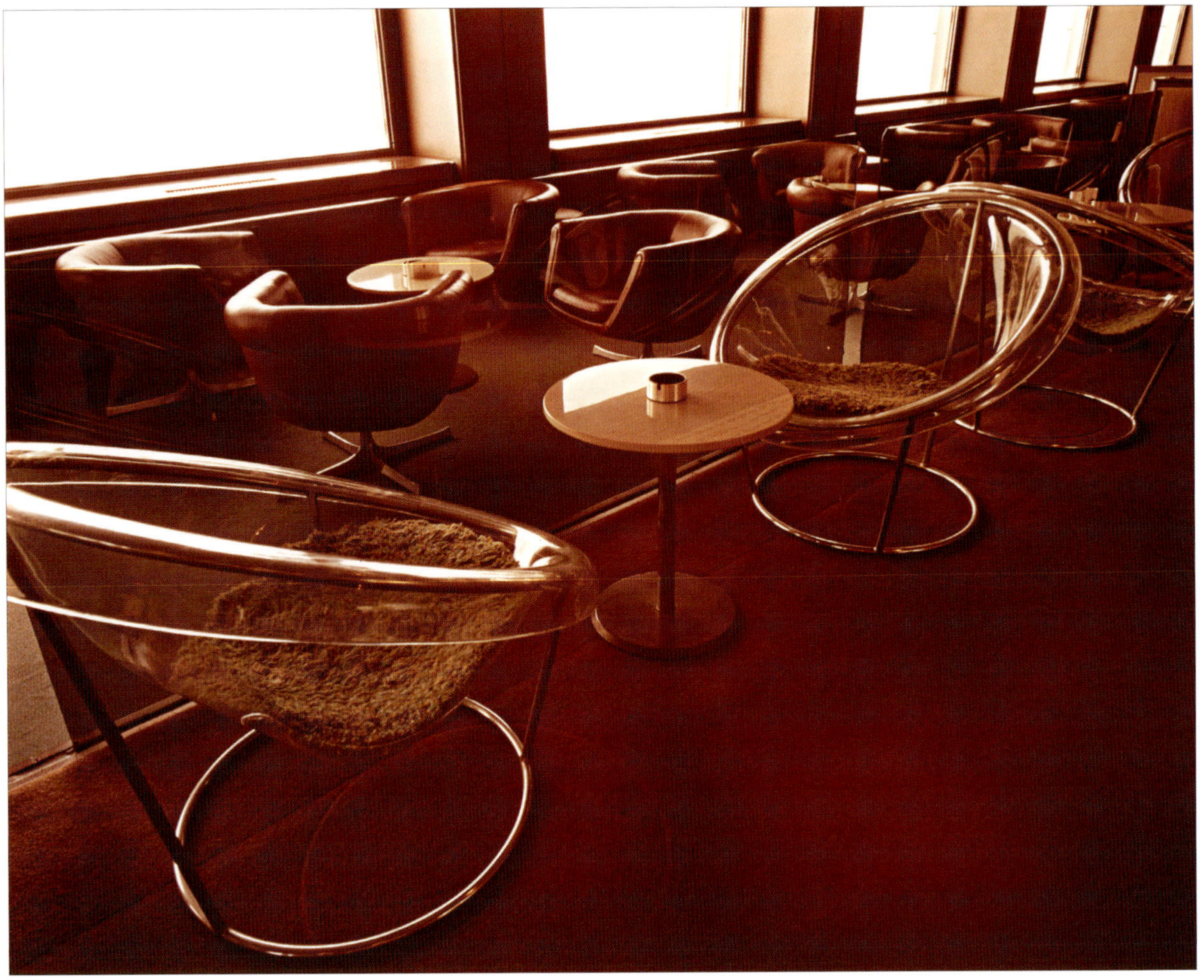

Kenmochi Design Associates, Interior of Tokyo Kaikan, 1968.

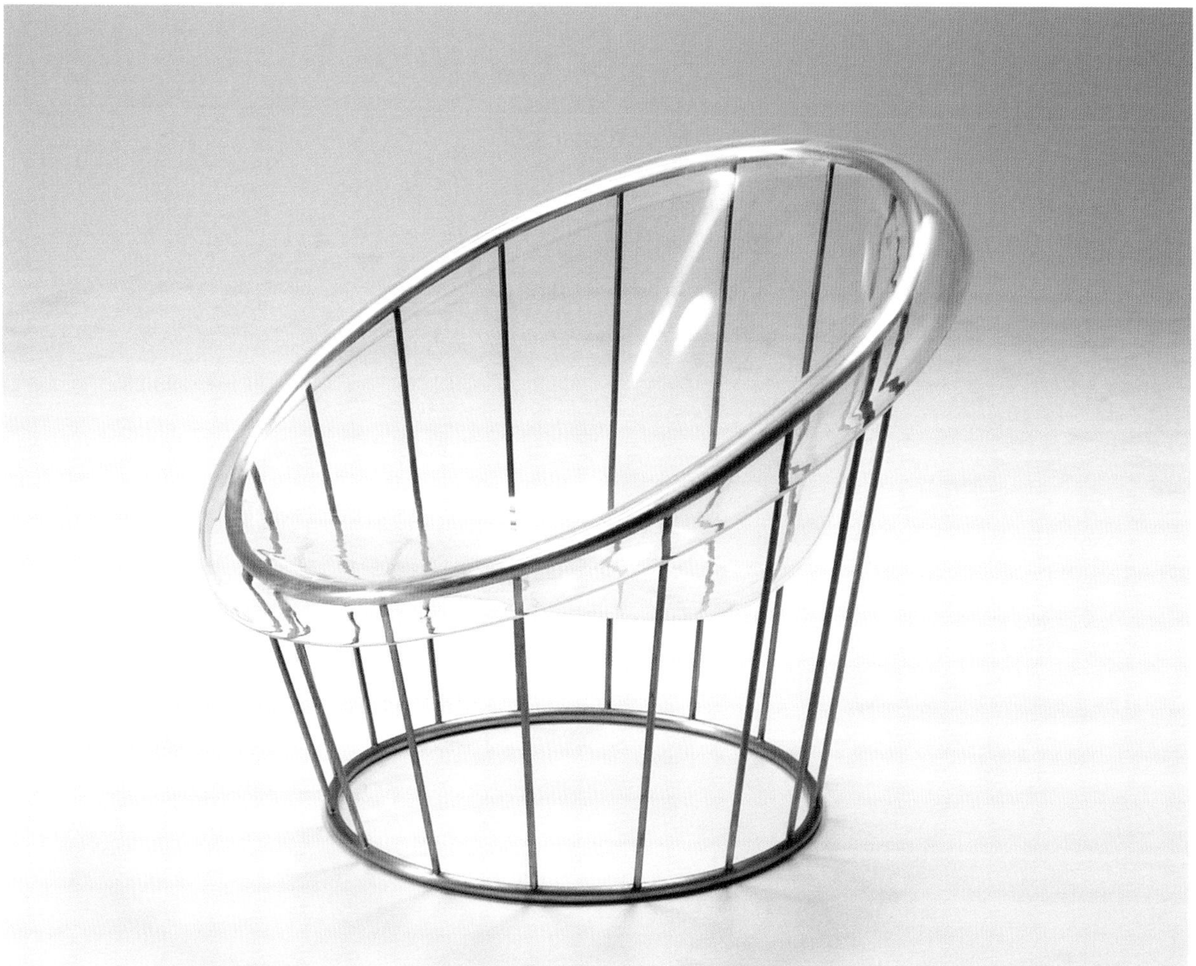

Kenmochi Design Associates, Easy Chair SM 7051, iron and acrylic, 1963 (reproduced 1969), 26 $3/4$ x 31 $1/2$ x 13 $3/8$ in. Original manufactured by Kyowa Gas Co., Ltd. Reproduction by Tendo Co., Ltd.

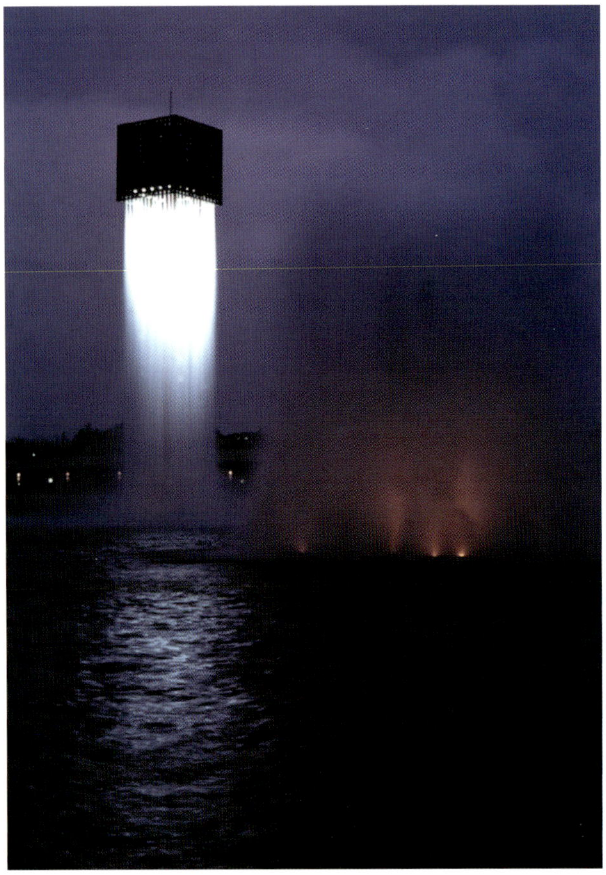

Isamu Noguchi, Expo '70 Fountains, Osaka World's Fair, 1970.

developed many more of Kenmochi's chair and industrial designs, and managed numerous significant interior-design commissions. Kenmochi's furniture designs may not seem as radical today as they were mid-century when the ongoing reality in Japan (until the economic boom of the 1980s) of an immature production system and a struggling economic market for modern furniture exports stifled his international ambitions. However, the products he made alone and in concert with the Industrial Arts Research Institute unequivocally advanced the Japanese design "identity" toward the international standard of furniture production.

Good design, good art, as Noguchi adamantly argued, should be available to everyone, to the global client. Noguchi's Akari lamps were manufactured on this premise, although today the economics of supply and demand have made his idealistic desires impracticable. Ironically, Akari lighting designs have had a fate similar to that of Kenmochi's rattan chairs, having been remade into unauthorized paper lamps that inundate chain furnishing stores in the United States. Noguchi's ubiquitous Glass-Topped Coffee Table, manufactured by Herman Miller, has also shown up in illicit subtle reincarnations in low-end furniture outlets since it was first made in 1944. These are the excesses of American postwar culture that Noguchi feared, where what is important, significant, and authentic is prone to lodge in our collective psyche as recycled and overexposed.

Around 1960 a deep pessimism began to appear in Kenmochi's writings. Japan was moving into a period of rapid economic growth, and the county was creating a new image for itself as a wealthy nation, increasingly asserting its presence in the world. While that spur of economic growth felt hopeful to some, Kenmochi's feelings of despair and emptiness seemed to increase. He wrote, "The fad for things Japanese is the same for Japanese design thinking, equivalent [to giving] foreign markets samurai topknots or kabuki . . . Japan is not a famous runner lined up at the hundred-meter mark. The solution for this is to give in to the real thing when it comes, and then, if you're brave, to absorb it, make it your own and release it again. It's no good if you don't have the real thing."[37]

Many influences that affected Isamu Noguchi and Isamu Kenmochi have not been mentioned in this very brief and general survey. However, Noguchi was asked once what he thought was the most important thing he acquired from his stay in Japan in 1950. His answer was "good friends." The friends Noguchi made during this trip were to remain close and supportive for the rest of his life. One of them was Isamu Kenmochi.

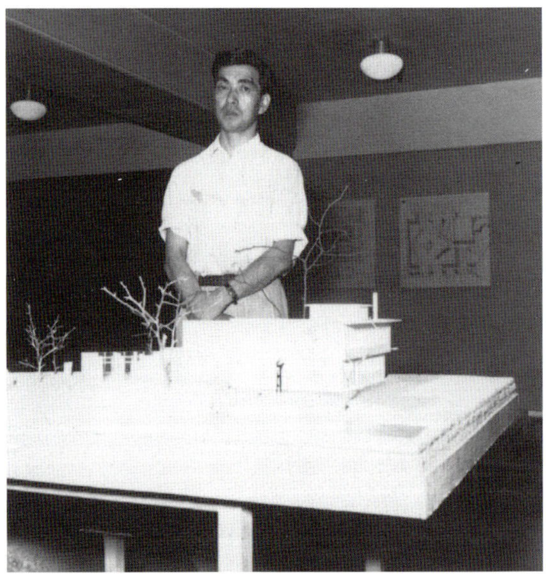

Yoshiro Taniguchi with a model of *Shin Banraisha*, Mitsukoshi Department Store exhibition, 1950.

Association with Noguchi and his methods as a sculptor gave Kenmochi much inspiration and amazement. He closely observed how Noguchi made plaster models of his creations before transforming them into full-scale, realized objects. Kenmochi felt that Noguchi's inventions had a feeling of Japanese taste that was dissimilar from what he had learned from Taut and Perriand. Noguchi's approach was more intuitive, and Taut and Perriand's approach, more studied. Kenmochi valued Noguchi's sensibility as a sculptor, not as a designer, and he observed first-hand Noguchi's process and awareness as a sculptor while collaborating with him on the Bamboo Basket Chair. The two Isamus remained close friends long after their collaboration in 1950–1952. Noguchi would make many introductions for Kenmochi when he visited the United States, including to fellow designers such as Charles and Ray Eames, Alexander Girard, Walter Gropius, George Nelson, and Mies van der Rohe. Meeting with these people had a profound impact on all of Kenmochi's subsequent industrial designs. Noguchi would go on to make only three distinct interior furniture designs after 1950: the granite table designed for the Chuo Koran Gallery in Tokyo in 1952; his celebrated Rocking Stool and Dining Table for Knoll Associates in 1955; and his Prismatic Table for Alcoa in 1957.

When Noguchi departed Japan on September 4, 1950, pleased with his new role as cultural emissary from abroad, he offered his friend, Kenmochi, a calligraphic drawing he had made of the kanji character for the word "courage." It is also the character for their shared name, Isamu.[38]

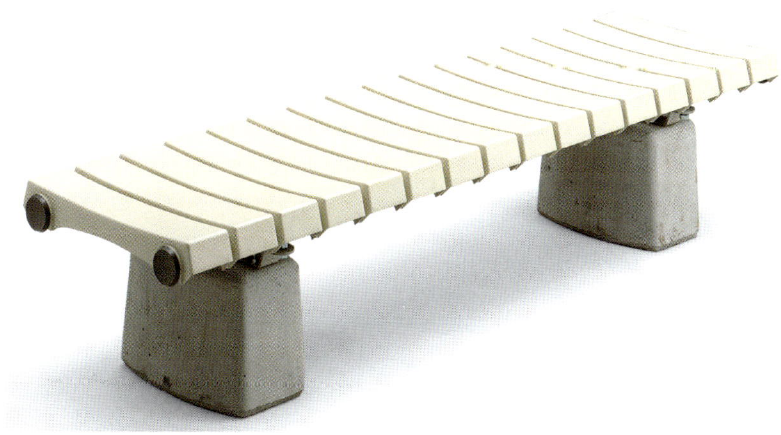

Kenmochi Design Associates, Bench for Expo '70, Osaka World's Fair, FRP, concrete, and iron, 1970. 14 15/16 x 18 7/8 x 87 7/16 in.

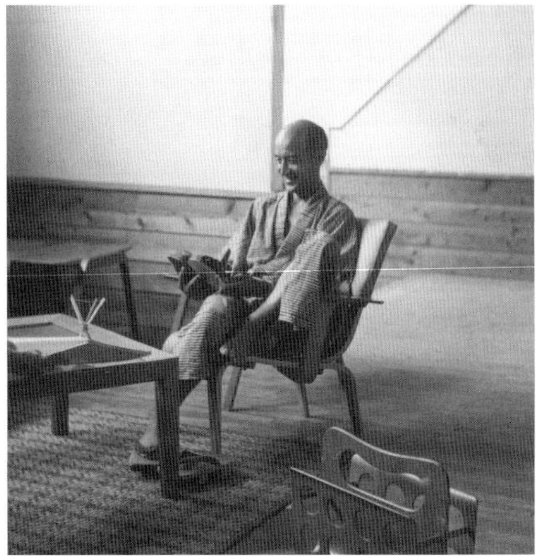

Isamu Noguchi wearing a striped *yukata*—a light summer kimono—during his visit to the Industrial Arts Research Institute, 1950.

NOTES

1. Masanao Kano, "Chairs and the Modern of Kenmochi Isamu," in *Japanese Modern: Retrospective Kenmochi Isamu*, ed. Hitoshi Mori (Matsudo: Matsudo City Cultural Foundation, 2004): 222.

2. Chairs for Kenmochi were a touchstone by which he could judge the efficiency of his attempts to realize the modern in tangible form. Arguing that Japan lacked a "science of chairs," he advocated chair designs based on the form of the seated human body. He put forth the prerequisite for "good Japanese chair prototypes based on the ideal of the sitting form as the outlet for the human body's shape." He pressed for an end to chairs and worktables that did not suit the physique of the "young industrial warrior." Kenmochi wrote several articles concerning the new state of design in Japan. Some of these include "Research Production: Molded Furniture: Molded Plywood Furniture," 1949; "Research Production: New Wood Products Made with High Frequencies," 1950; "Contemporary Living and Dining Rooms," 1950; "Design and Designers," 1951; and his most famous essay, "Japanese Modern and Japonica Style," from 1954. Ibid., 219–222.

3. Ibid., 219.

4. Regular bouts of self-loathing and doubt, interspersed with manic moments of extreme self-adulation, marked the life of Kenmochi from the time he was a young man. Many of these episodes affected his busy work schedule. It is not known what prompted his suicide, although there has been speculation that he was despondent because he felt that a new generation of designers was overlooking and overtaking his accomplishments. Hitoshi Mori, "Tradition and Modernity: Building Bridges and Interstices: Notes on Kenmochi Isamu," in *Japanese Modern: Retrospective Kenmochi Isamu*, 233.

5. The extent of Taut's knowledge of Japanese architecture before he visited Japan is unknown. Taut traveled to Japan via Siberia directly to Kyoto, arriving in 1933. On his second day in the country he was taken to the Katsura Imperial Villa by a group of architects from Osaka calling themselves the Japan International Architectural Association. The group's leader was Isaburo Ueno (1892–1972), who had studied at the Wiener Werkbund in Germany and embraced Taut's philosophy. With Katsura under the strict administration of the Imperial Household, for a non-Japanese to be taken to visit such an important symbol of imperial authority was quite an audacious move. Arata Isozaki, *Japan-ness in Architecture* (Cambridge, MA: MIT Press, 2006): 12–13.

6. The Bauhaus was the art and design school—there was no official architecture department at the beginning—that operated in Germany from 1919 to 1933, and then in Chicago from 1937 to 1938. It had four different directors: the founder, Walter Gropius, from 1919 to 1928; Hannes Meyer, from 1928 to 1930; Ludwig Mies van der Rohe, from 1930 to 1933; and Laszlo Moholy-Nagy, from 1937 to 1938.

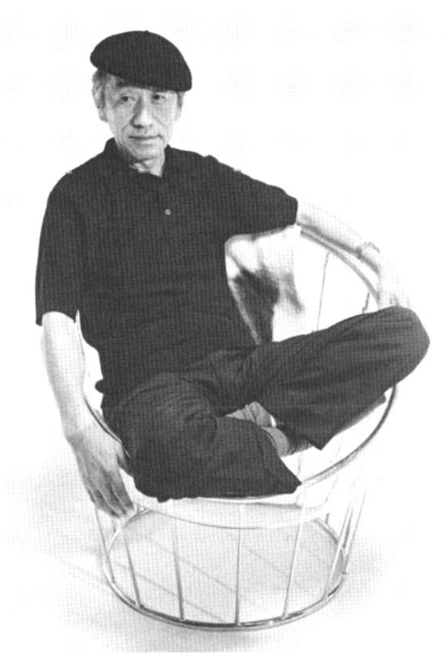

Isamu Kenmochi seated on his Easy Chair SM 7051, 1963.

7. Hitoshi Mori, "Tradition and Modernity: Building Bridges and Interstices: Notes on Kenmochi Isamu," in *Japanese Modern: Retrospective Kenmochi Isamu*, 234.

8. Kano, "Chairs and the Modern of Kenmochi Isamu," in *Japanese Modern: Retrospective Kenmochi Isamu*, 220.

9. Yuko Hashimoto, "A Brief History of the Foundation," *Guidebook: Photo Archive of the Japan Industrial Arts Foundation* (Tokyo: Industrial Arts Foundation, 2004): 6.

10. Kano, "Chairs and the Modern of Kenmochi Isamu," in *Japanese Modern: Retrospective Kenmochi Isamu*, 219.

11. Junzo Sakakura (1901–1969) also has links to Isamu Noguchi. Sakakura won the commission to build the Museum of Modern Art in Kamakura, selected from among five architects—including Kunio Maekawa, Junzo Yoshimura, Toshio Yamashita, and Noguchi's friend and collaborator, Yoshiro Taniguchi. Completed in 1951, the museum is a two-story, steel-framed structure based on the design of Le Corbusier's "Museum of Unlimited Growth" concept, published in 1939. In echoing Le Corbusier's system for domestic architecture from the 1920s, Sakakura also emphasized a harmony between the building and its natural environment. Built over a pond, the museum makes use of a uniquely Japanese tradition in its incorporation of the Japanese "oya" stone, a malleable soft stone used widely in Japan for constructing walls, moats, and storehouses. It was also used as a decorative material by Frank Lloyd Wright in the Imperial Hotel. Noguchi would have his first one-person museum exhibition at the Kamakura Museum of Modern Art in 1952.

12. Mary McLeod, ed., *Charlotte Perriand: An Art of Living* (New York: Harry N. Abrams, in association with The Architectural Leagues of New York, 2003): 90–91.

13. Isozaki, *Japan-ness in Architecture*, 9.

14. The Imperial Hotel did not appear "Japanese" to the Japanese (though it did to Western viewers), lacking the flatness and perspective particular to Japanese space. The design was more conventionally Beaux-Arts, comprising layers of planes without regard to the Japanese sense of graduated perspective. Wright drew much of his knowledge of Japanese architecture from the replica of the Phoenix Hall in Uji, Kyoto, which he saw at the World's Columbian Exposition of 1893 in Chicago. Tenshin Okakura's *Book of Tea*, from 1922, also helped define "Japan-ness" for Japan and the West. Isozaki, *Japan-ness in Architecture*, 8–10.

15. Ibid., 9–10, 14.

16. Kano, "Chairs and the Modern of Kenmochi Isamu" in *Japanese Modern: Retrospective Kenmochi Isamu*, 220.

17. The Bollingen Foundation was established by Paul and Mary Conover Mellon in 1945 to support scholarship in the humanities and social sciences with research fellowships, to be followed by the publication of the research material under the Bollingen Series imprint. Named for the small village in Switzerland where Carl Gustav Jung, father of analytical psychology, had his personal retreat, the foundation was initially established to assure wider access to Jung's work by English-speaking audiences. During its nineteen years of activity, the Bollingen Foundation provided financial assistance to many scholars and intellectuals.

18. Bert Winther, *Isamu Noguchi: Conflicts of Japanese Culture in the Early Postwar Years* (Ph.D. dissertation, New York University and UMI Dissertation Services, 1992): 154.

19. Ibid., 156.

20. Yonejiro Noguchi (1875–1947) was the first Japanese-speaking poet to be published in English. Before traveling to San Francisco in 1893, he studied at Keio University, where he met the poet Joaquin Miller. It was through Miller that Noguchi met Edwin Markham, Charles Warren Stoddard, and Gelett Burgess, among many others. He left San Francisco in 1900 for the East Coast, where he met Leonie Gilmour, Isamu Noguchi's mother. She helped the older Noguchi, editing his most famous pieces of fiction, including *The American Diary of a Japanese Girl.* He subsequently went on to England where he met William Butler Yeats, Thomas Hardy, and Ezra Pound. In 1904 he returned to Japan via the United States.

21. Shigeru Yura, "Collaboration for Monumentality," in *Banraisha: A Poetic Architecture by Yoshiro Taniguchi and Isamu Noguchi* (Tokyo: Kajima Institute Publishing Co., Ltd., 2006): 51.

22. In 1938, like so many of his colleagues in the arts, Taniguchi traveled to Europe, ostensibly because of a commission to design a garden in Berlin for the Japanese Embassy. With war rapidly approaching, Taniguchi decided to forego the commission and travel in Germany, France, and Italy, ultimately making his way back to Japan via Denmark and the United States. Notable examples of Taniguchi's work include Akasaka Palace, the Imperial Theater, the Asian Gallery at the Tokyo National Museum, and the Hotel Okura, across from the American Embassy in Tokyo. Makiko Sugiyama, "Yoshiro Taniguchi and Keio University," in *Banraisha: A Poetic Architecture by Yoshiro Taniguchi and Isamu Noguchi*, 87–89.

23. Winther, *Isamu Noguchi: Conflicts of Japanese Culture in the Early Postwar Years*, 130, 141.

24. Tange was able to combine Modernist and nationalist themes. Isozaki, *Japan-ness in Architecture*, 15–17.

25. Isozaki, *Japan-ness in Architecture*, p. 15.

26. "Japanism," a term that originated in France as *Japonisme*, an original French term, refers to the influence of the arts of Japan on the West. Japanese woodblock prints (ukiyo-e) were a major influence, and provided a source of inspiration for European Impressionist painters and also influenced practitioners of Art Nouveau and Cubism. Works marked by Japanism included a lack of perspective, flat areas of color, off-center subject matter, and low diagonal axes in the background. The architect Frank Lloyd Wright is known to have collected ukiyo-e prints. See Isozaki, *Japan-ness in Architecture*, pp. 8–10.

27. Winther, *Isamu Noguchi: Conflicts of Japanese Culture in the Early Postwar Years*, 142.

28. Kazuyoshi Kudo in conversation with Bert Winther, August 1991. Kudo wrote *Japanese Baskets* (New York: Kodansha International, 1980). Ibid., 153, note 68.

29. One of the greatest disappointments Noguchi experienced was his rejection as the designer for the memorial in Hiroshima's Peace Park, and he took this loss as a personal affront. Additionally, the Bell Tower was never realized, and only the commission for bridge railing connecting Peace Park to the city was constructed.

30. On view at Mitsukoshi was the full-scale plaster model for the sculpture *Mu*, which would be remade in sandstone and installed outdoors the following year at Keio. Other objects included the Seto ceramics, and furniture and other sculptures destined for Keio. Additionally there were photographs of previous projects and sculptures that illustrated Noguchi's

international standing as an artist. Prominently featured were the two models for the Hiroshima Bell Tower. Winther, *Isamu Noguchi: Conflicts of Japanese Culture in the Early Postwar Years*, 145–146.

31. Ibid., 133–134.

32. Dore Ashton, *Noguchi: East and West* (New York: Alfred A. Knopf, 1992): 96.

33. Noguchi was well acquainted with the idea of the "primitive" much earlier than his discovery of the ancient Jomon art of Japan. Studies of European and contemporary art in America in the 1920s were filled with images of Picasso mining African art, and Henri Matisse, Paul Klee, and Constantin Brancusi unearthing artifacts from other cultures for their own purposes. The results of this engagement were referred to as "primitive." When Noguchi traveled to Japan in 1930, he visited the museums of Tokyo and Kyoto, which sponsored research on *haniwa* and the large tumuli of the fifth through seventh centuries. His sculpture *The Queen*, from 1931, was initially published under the title *Haniwa* and bears a close resemblance to *haniwa* forms from the sixth century. Noguchi may have turned to the *haniwa* form because it "allowed him to create a symbolic authority that was of greater antiquity and at the same time of more advanced contemporaneity" than the Japanese culture that existed at that time. Like many artists, Japanese and others, who came upon this ancient form and appropriated it, Noguchi felt that this old work seemed more new than the present-day Japan he was experiencing. Bert Winther-Tamaki, "The Ceramic Art of Isamu Noguchi: A Close Embrace of the Earth," in *Noguchi and Modern Japanese Ceramics: A Close Embrace of the Earth* (Washington D.C.: The Arthur M. Sackler Gallery, Smithsonian Institution, in association with University of California Press, 2003): 15–17.

34. Kenzo Tange was undoubtedly influenced by Okamoto's formulation of Jomon fundamentalism. In 1956, Okamoto created a series of ceramic tile murals for Tange's new Tokyo City. Winther, *Isamu Noguchi: Conflicts of Japanese Culture in the Early Postwar Years*, 117.

35. Ibid., 117.

36. Isozaki, *Japan-ness in Architecture*, 4.

37. Quoted in Kano, "Chairs and the Modern of Kenmochi Isamu," in *Japanese Modern: Retrospective Kenmochi Isamu*, 222.

38. Winther, *Isamu Noguchi: Conflicts of Japanese Culture in the Early Postwar Years*, 159–160.

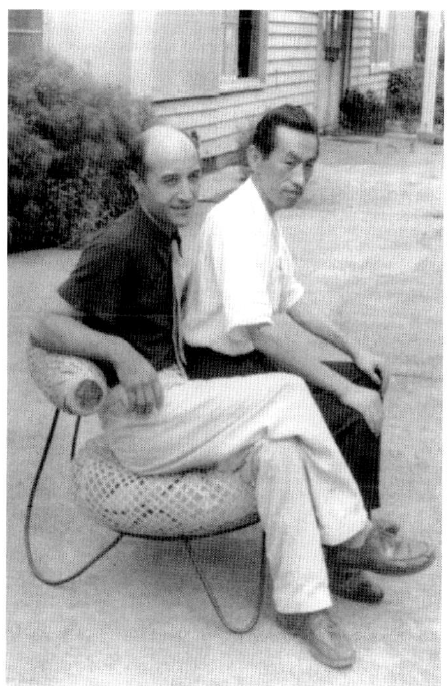

Isamu Noguchi and Isamu Kenmochi, seated on their Bamboo Basket Chair, 1950.

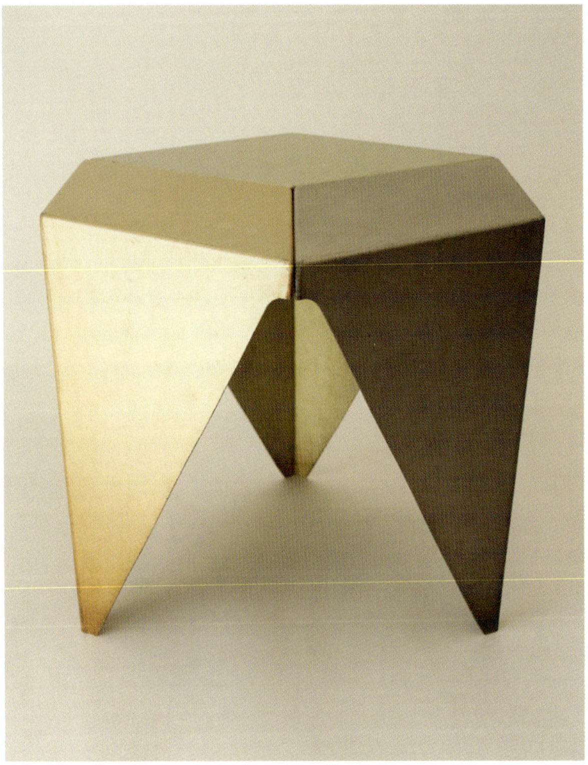 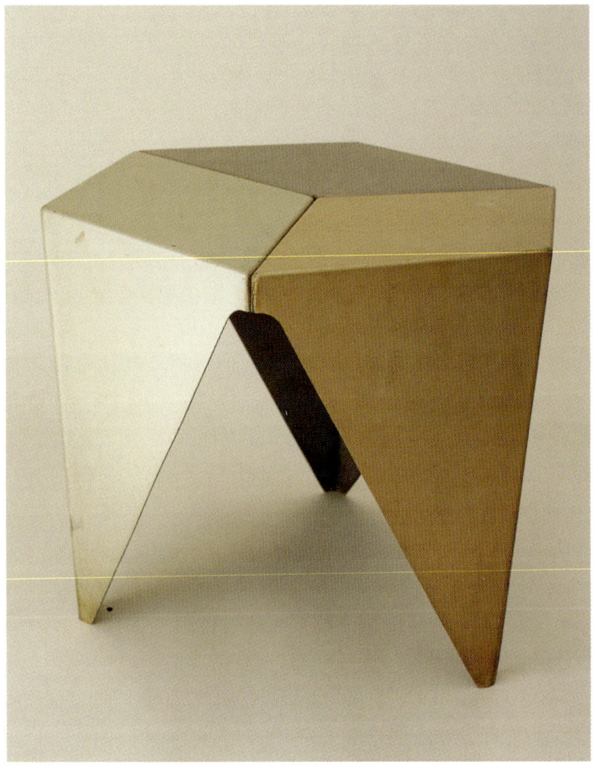

Above and opposite top: Isamu Noguchi, Prismatic Table, anodized aluminum, 1957, 16 x 17 x 19 in. Noguchi was one of twenty designers invited by the Aluminum Company of America (Alcoa) to create modern aluminum furniture designs for the company's Forecast Program. The program hoped to "forecast" the innovative potential of aluminum in industrial and home furnishing, and included the participation of renowned designers such as Herbert Bayer, Charles Eames, and Garret Eckbo. Alcoa projected that Noguchi's table could be produced for as little as thirteen dollars, although ultimately it was never mass-produced.

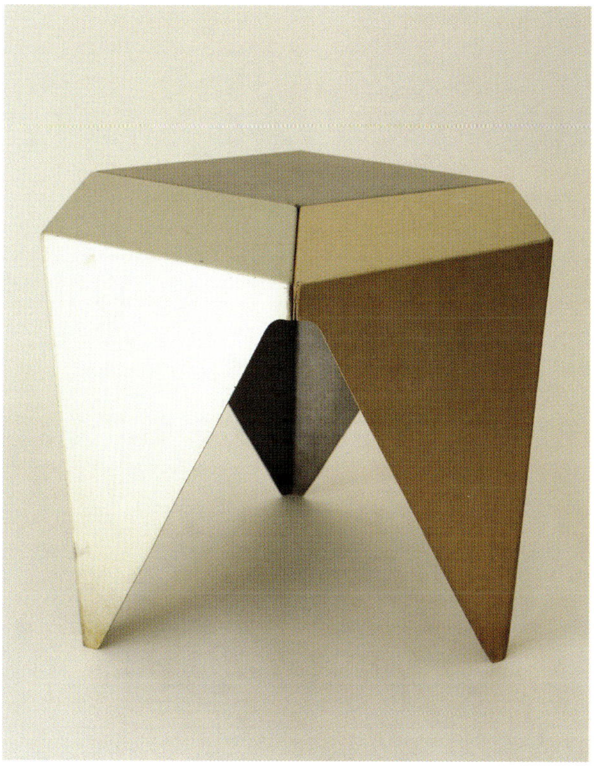

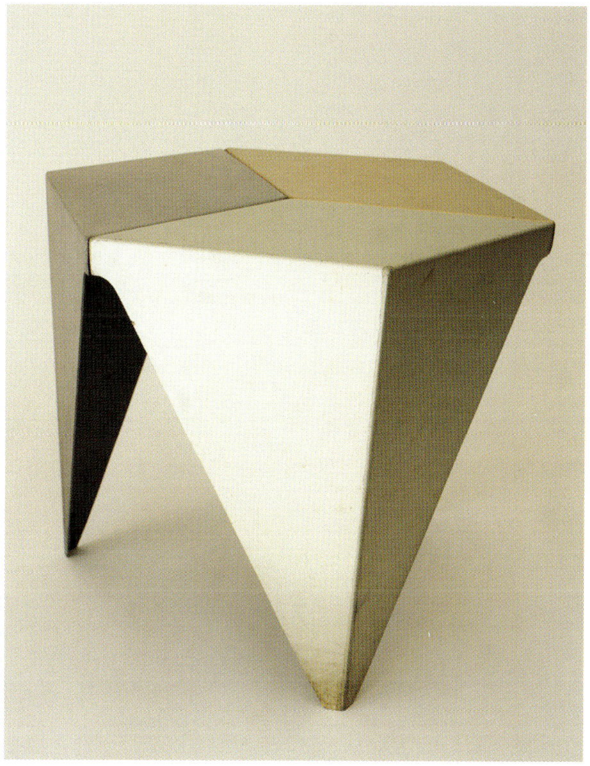

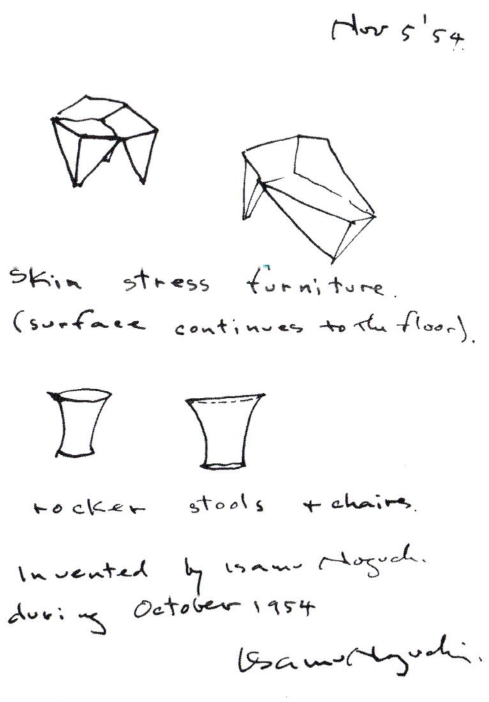

Above: Isamu Noguchi, Skin Stress Furniture, ink on paper, 1954, 11 x 8 7/16 in.

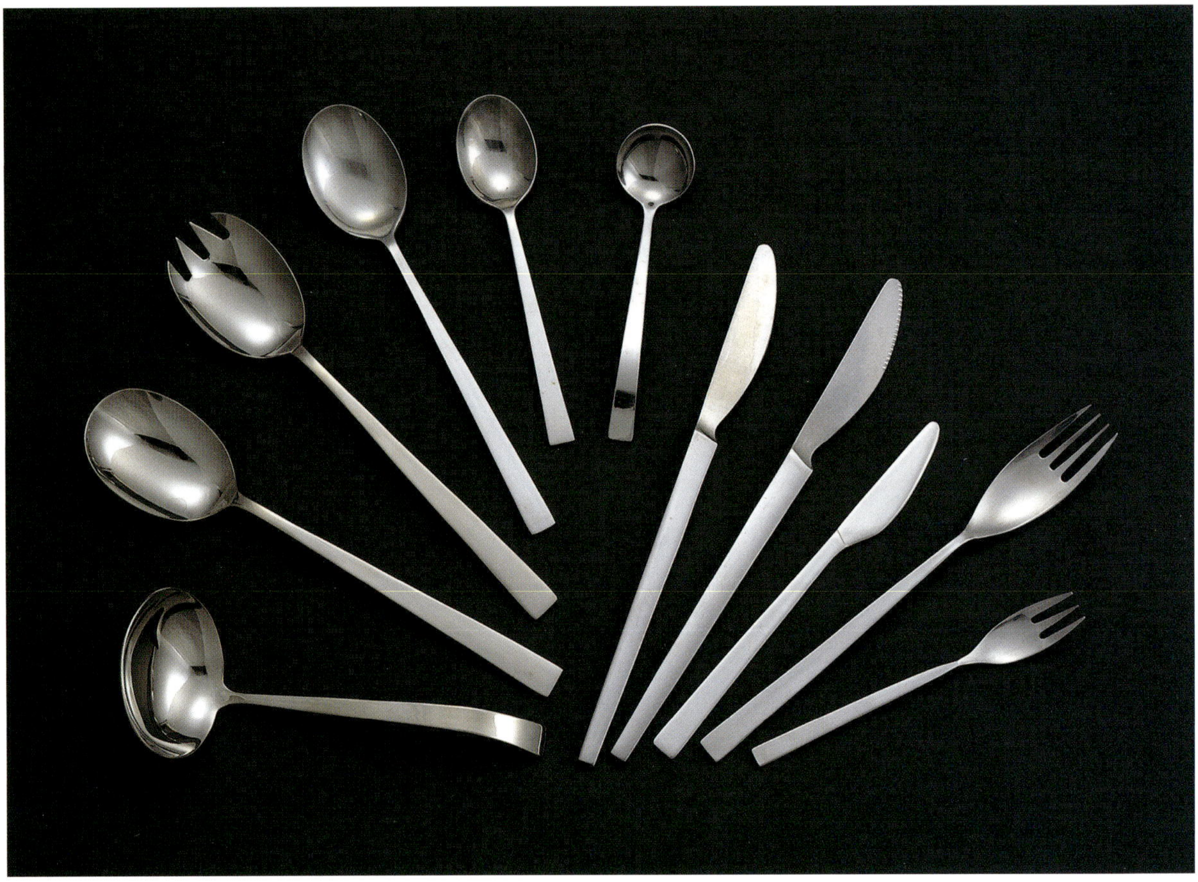

Kenmochi Design Associates, Utensils "SASA," stainless steel, 1964; meat knife: 8 7/16 in.; fish knife: 8 7/16 in.; butter knife: 6 in.; serving spoon and serving fork: 7 7/8 in.; teaspoon: 5 3/8 in.; fork: 7 11/16 in.; soup spoon: 7 3/16 in.; small serving spoon: 4 11/16 in.; small fork: 5 1/8 in.; ladle: 6 7/16 in. Manufactured by Sato Shoji.

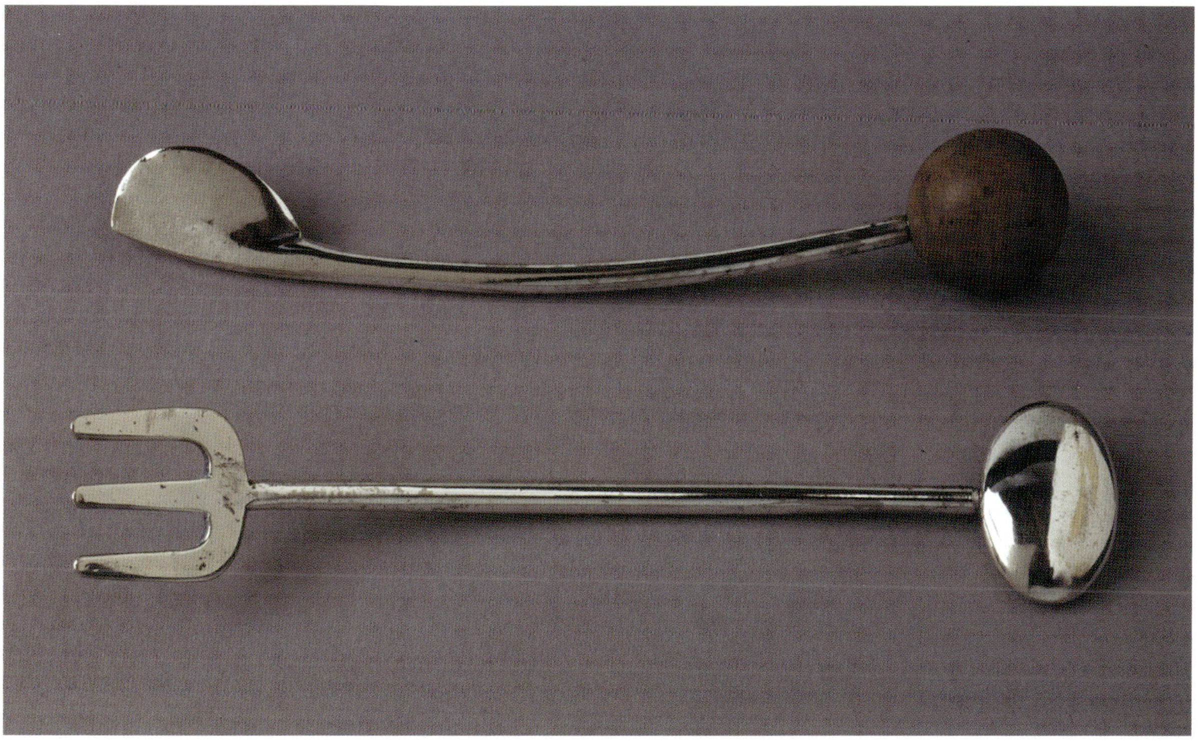

Isamu Noguchi, Knife and Fork, sterling and wood, ca. 1960; knife: 5 x 7/8 x 7/8 in.; fork: 6 1/16 x 11/16 x 1 in.

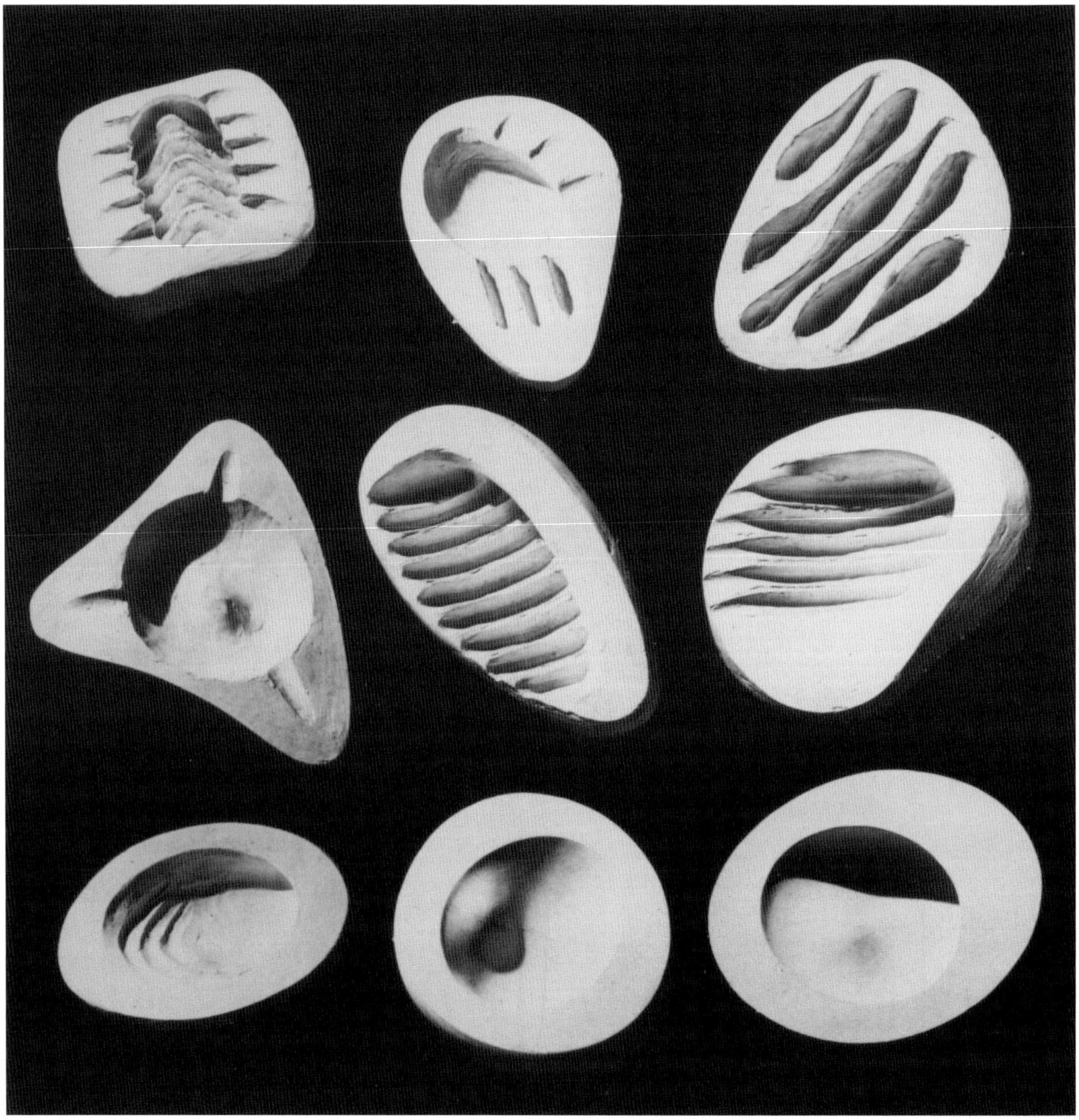

Above: Isamu Noguchi, Nine Ideas for Ashtrays, 1948–1956. Opposite: Isamu Noguchi, Ashtrays, ca. 1945. These illustrations were designed for an unpublished magazine article, "The Sculptor and the Ashtray," text by Mary Nix, ca. 1945. Noguchi was a smoker for many years, as were many of his contemporaries. In the 1940s ashtrays were ubiquitous, and interest in creating a well-designed ashtray was implicit for any industrial designer. Noguchi seems to have taken the notion of designing the perfect ashtray seriously, and went so far as to patent three different designs. A mock-up for an article exists that illustrates various ashtray designs, and was to include an article by Mary Nix on what she described as, "A story of two ashtrays. It is also the story of the man who made them and why his designs turned out the way they did. Incidentally, it is the story of all the ashtrays you ever bought, saw, or use."

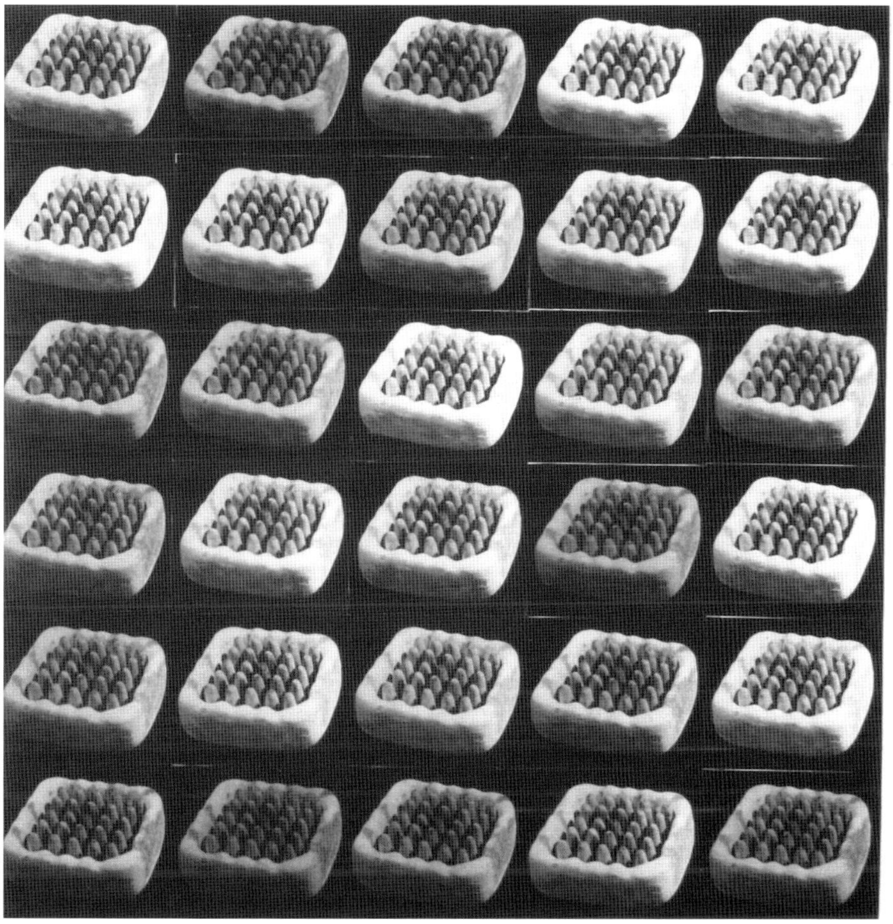

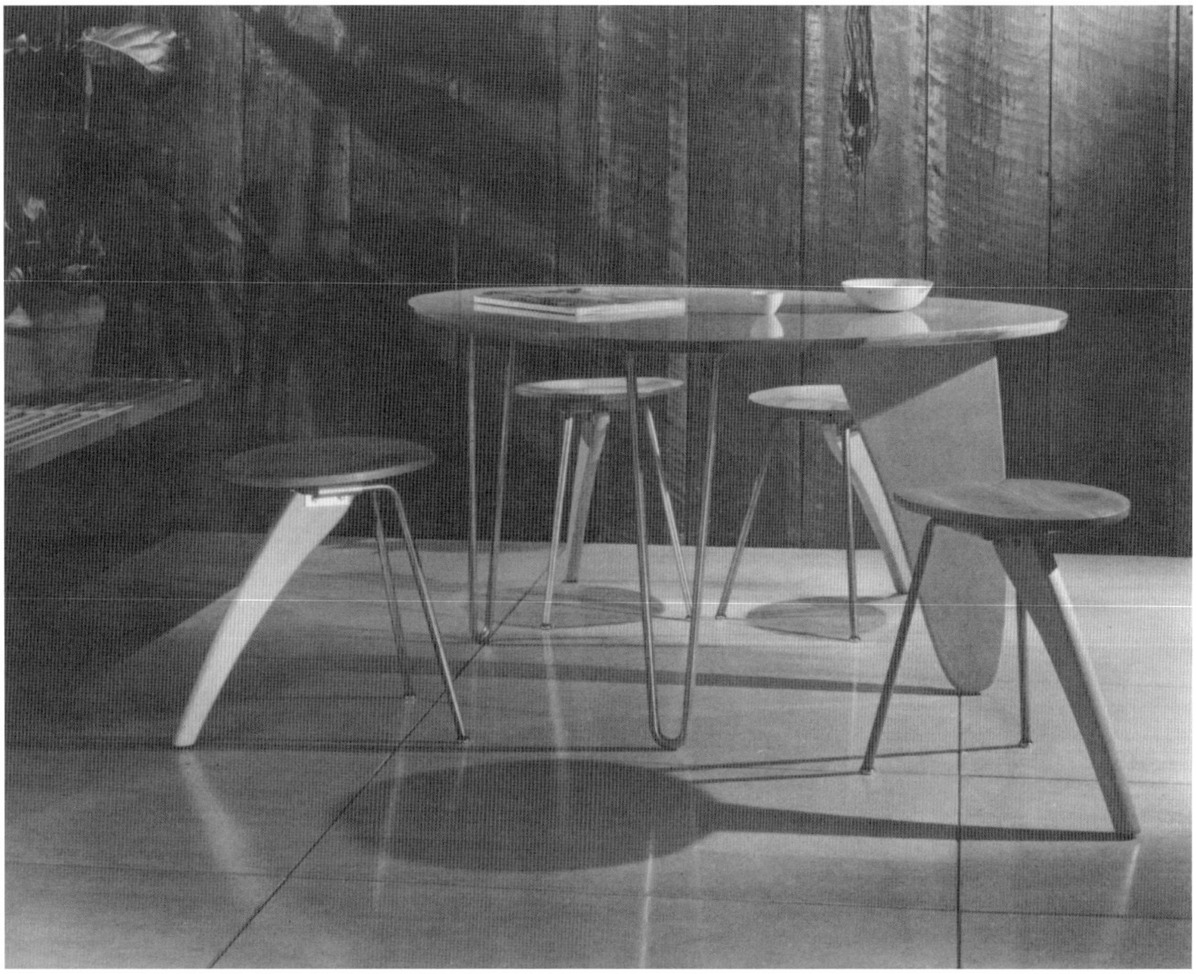

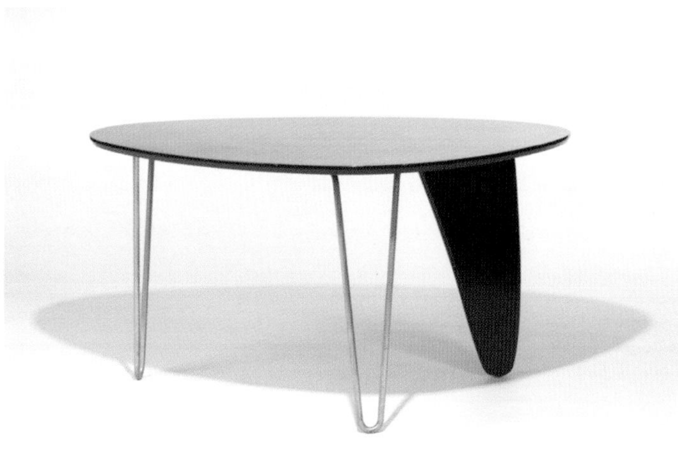

Top: Isamu Noguchi, Dining Table and Chairs, 1949–1950. Table: birch and steel rods, 26 x 50 x 35 in.; chairs: birch, steel, and rubber, 17 x 14 x 19 in. Bottom: Coffee Table, ebonized birch and steel rods, ca. 1949–1950, 25 1/2 x 50 x 36 in.

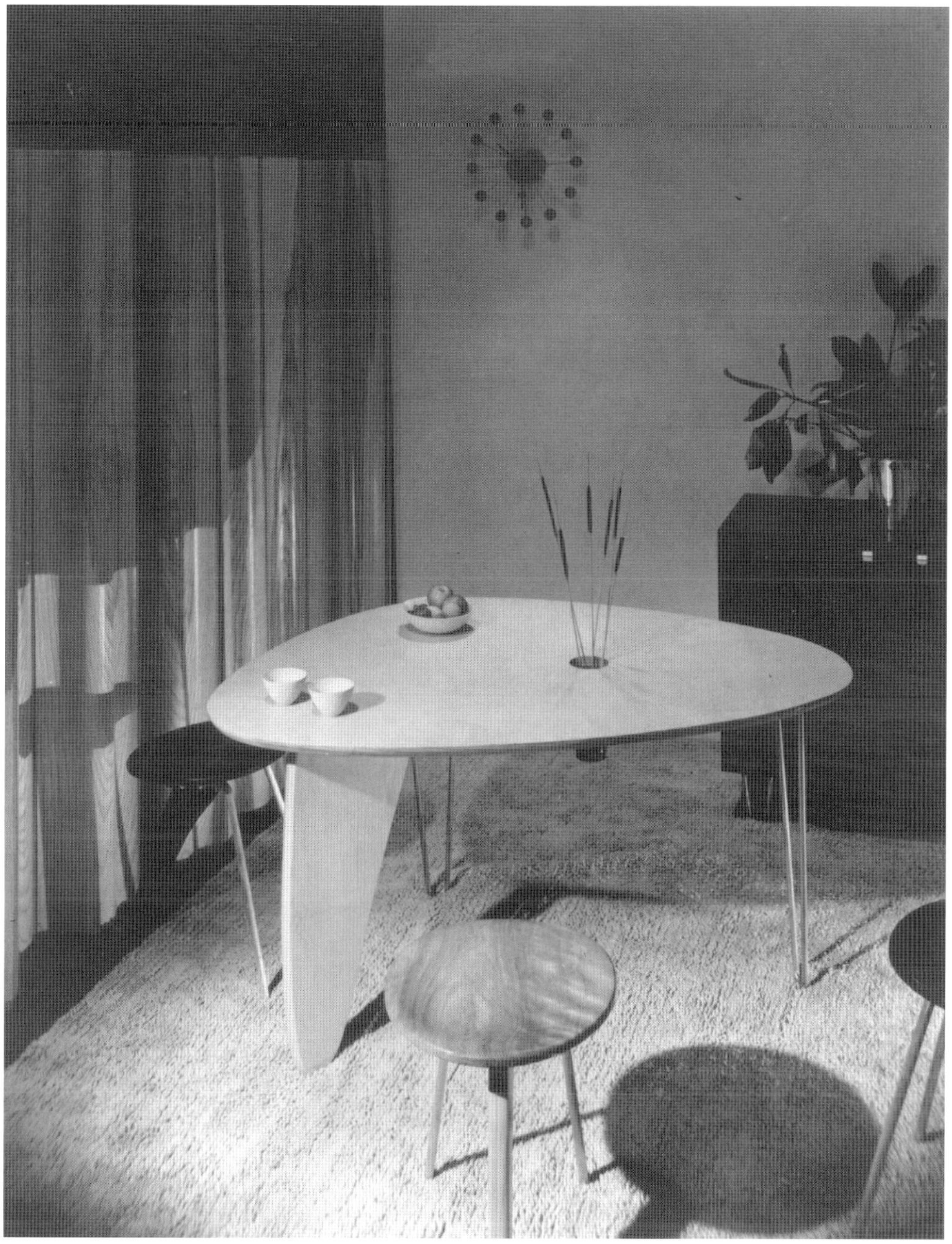

Unknown photographer, Living Room with Noguchi's Furniture, ca. 1950. The vital formal language of biomorphism survived through World War II, as witnessed by this dining table and stools and a companion coffee table (nicknamed the Rudder or Fin Table and Stool) in which Noguchi ignored the conventional composition of a table as a rectangle on four legs. Instead, he chose curvilinear shapes and set up an amusing contrast between the rounded triangle of the top and the elongated triangles of the three legs. A hole in the top of the dining table allows for the insertion of a bowl. This engaging detail and the play between open and closed forms relates to Noguchi's Marble-Topped Coffee Table of 1948.

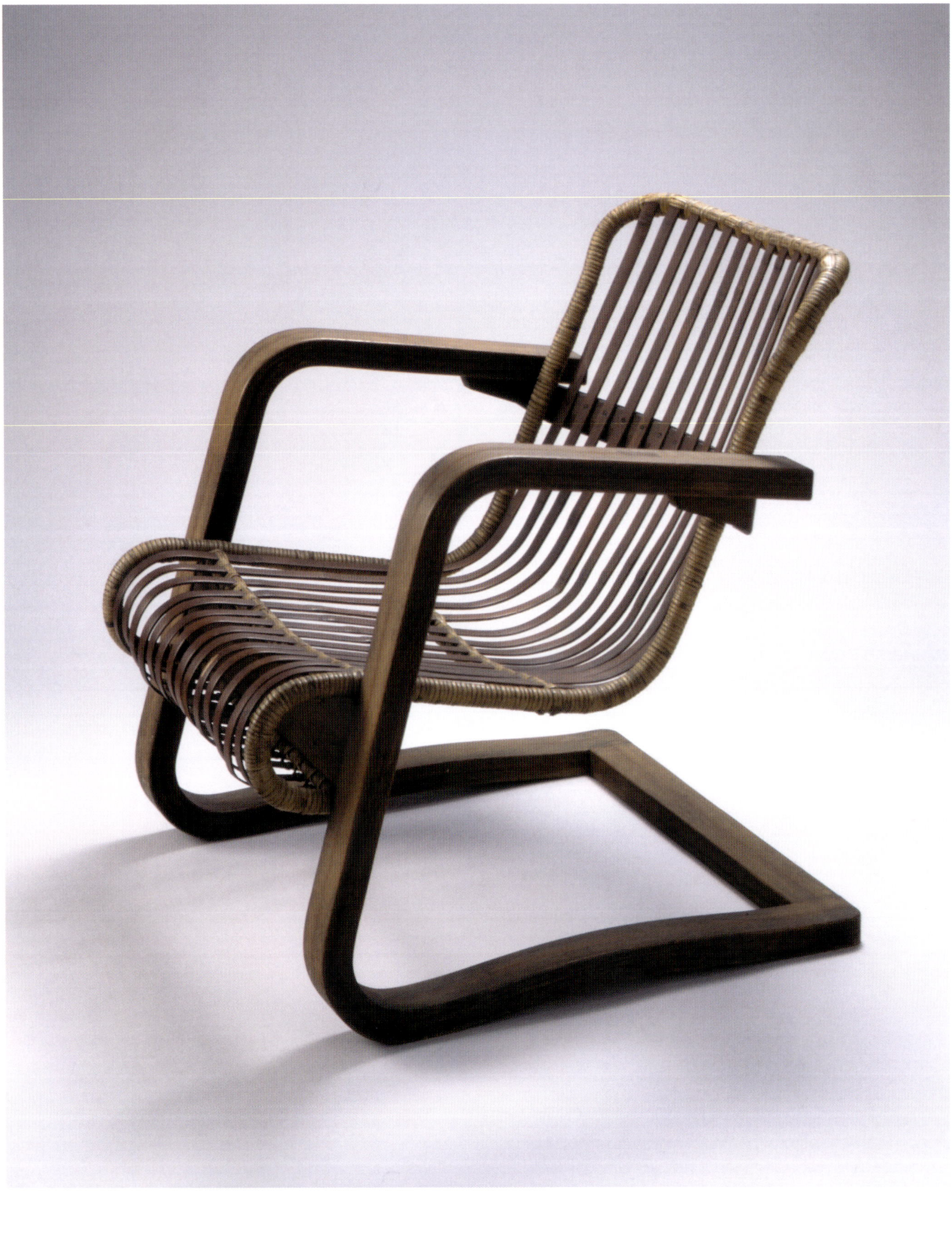

Industrial Arts Research Institute, Bamboo Chair Prototype, ca. 1940s.

ISAMU KENMOCHI AND ISAMU NOGUCHI:

THE MEETING AND ITS MEANING

HITOSHI MORI

Although unplanned, Isamu Noguchi was asked to take part in the creation of *Shin Banraisha* in 1950, when he returned to Japan after sixteen years. And as luck would have it, Isamu Kenmochi, a designer at the Industrial Arts Research Institute, became involved in this project. In postwar Japan, Noguchi's visit, along with the works he produced, exerted great influence on the field of art and design. At the same time, through Noguchi's long-held motivation to create and Kenmochi's desire to develop the country's design industry, each gained new possibilities through their encounter with each other. Here, we look into the meaning of their collaboration and the effect it had on designers in Japan. We focus on the significance of their meeting, in particular the way in which it triggered a fundamental question: the eternal theme of tradition versus creation.

An Inevitable Chance Encounter

Noguchi and Kenmochi first met on June 24, 1950, at Kenzo Tange's office at the Tokyo University Faculty of Engineering.[1] It is safe to assume that Kenmochi knew Noguchi's name as a designer; though Kenmochi was in Sendai after the war, he was introduced to the quality of American furniture through his furniture production for the United States General Headquarters, which led to his research on modern designs by companies such as Herman Miller. Shortly after their first meeting, Kenmochi was asked by their mutual friend, Genichiro Inokuma, to find a studio space for Noguchi. The three men shared a strong awareness that art should have a larger role in society. At that time, Noguchi was on his world tour, pursuing the theme of "leisure" through a grant from the Bollingen Foundation. His goal was to "contribute to society by relating . . . arts and community," clues for which he attempted to find in historic sites around the world.[2] To that end, a proposal to create playgrounds had also been made, but was not yet realized. Noguchi's notion that "sculpture and architecture should have a more comprehensive relationship" was confirmed by "the relationship between Japanese architecture and the garden, the unison of human living and nature."[3] Moreover, the proposed concept for *Shin Banraisha* was just that—to unify garden and interior spaces—and Noguchi accepted the offer from Yoshiro Taniguchi to co-design the project.

On July 28, following a rainy day, Noguchi visited the Industrial Arts Research Institute wearing Inokuma's *yukata* (a light summer kimono). He liked the place so much that he set to work immediately on August 1. Kenmochi, then First Chief of Research, treated Noguchi's work as a new experiment for the Institute, and arranged to give it the status of an official project. He donated not only the space, but materials and tools as necessary. Noguchi requested Tsutomu Hiroi as his assistant, but in fact the entire staff of the Institute was made available for his projects. Works produced during this period include sculptural maquettes for *Wakai Hito*, *After Heroism*, *Bell Tower for Hiroshima*, and others; the sculpture *Mu*; and furniture, including Coffee Table, Stool, and Bamboo Basket Chair.

The wood maquettes for metal sculptures were finished in rust lacquer to achieve a metallic texture. The subject and style of these maquettes were similar to Noguchi's sculptures from the 1940s, the group that included

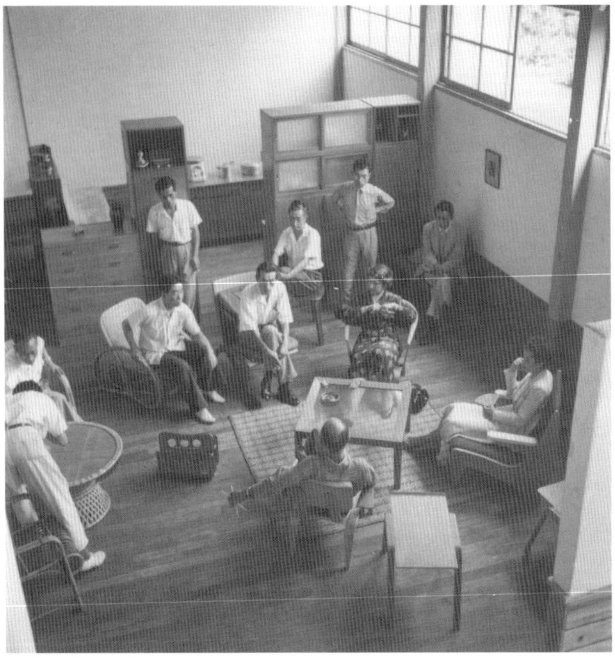

Isamu Noguchi (foreground, with back to camera) with Isamu Kenmochi, Genichiro and Fumiko Inokuma, and staff at the Industrial Arts Research Institute, July 28, 1950.

Keitai (1945). However, the remaining pieces were altogether distinguishable, conceived from his experience during this stay in Japan.

Mu was inspired by the shape of a *tourou*, a traditional outdoor lantern. Although he consulted with Inokuma, the schedule did not allow its production in stone. Refusing to give it up entirely, Noguchi began working in plaster on August 8, and managed to complete the piece in a mere two days.[4] A late-night photograph captured Noguchi at work, unable to pause from the production.

The design of the stool avoids symmetry, as seen in the ancient Chinese *shaku*, a three-legged wine vessel. For the table that accompanied the stool, a free-form top was supported by legs designed to expose the connection. The two pieces of furniture were drawn at full scale by Noguchi, and were then fabricated by the engineers at the Institute.

Noguchi came up with an idea for seating using "the thread construction of a Japanese *tsuzumi* (drum)," and brought to the Institute a simple model made with "four chopsticks tied into an X-shape."[5] It was a stool with two pieces of plywood bound by rope; however, plywood technology at that time was not advanced enough to produce it.[6] Kenmochi showed Noguchi some plywood furniture, bamboo furniture, and lacquered metal tableware, all made at the Institute, and asked for his critique. Noguchi showed interest in the thick long arms that extended from a Chinese round-back chair. But as soon as Kenmochi showed his long-held idea of making a chair from a woven basket, Noguchi was quickly taken by it. "This is very interesting," he said. "Let's make this chair. Let's design it together." He altered his plans and came the next day with a sketch for it. In Kenmochi's prototype the chair had wooden legs, but Noguchi modified it to incorporate "a long steel rod bent to form a gentle curve." To this Kenmochi reacted that "it was so original, my first shape could not even be compared with it."[7] Based on Noguchi's sketch, Kenmochi made a full-scale detail drawing, followed by a model in wire and clay so that the two of them could study the shape. The curve of the leg was adjusted, and the banana-shaped back was made thicker. The blueprints could now be made from the revised model. The steel frame was fabricated by the Institute's metal shop, while the bamboo elements were made by technician Harumitsu Inoue. As the piece was not approved until a very late stage of production, only half a day was given to finishing the chair. This could be the reason that the official cameraman of the Institute, Masao Usui, recorded all other aspects of production but this one. The morning the chair was completed and was ready to be sent out, Noguchi sat on the chair and said with satisfaction, "Not bad!" A photograph

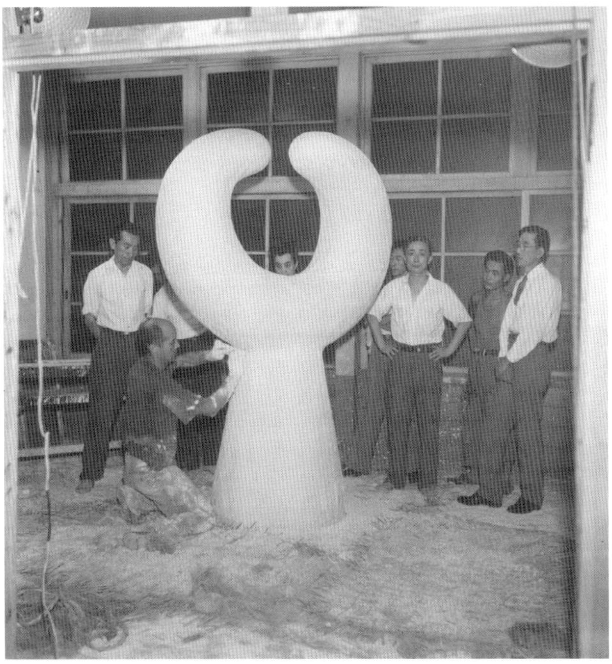

Isamu Kenmochi and other Industrial Arts Research Institute staff watch Isamu Noguchi at work on *Mu* (Nothingness), August 10, 1950.

of the two of them sitting together on the chair shows contentment, with a glimpse of nervous anticipation.

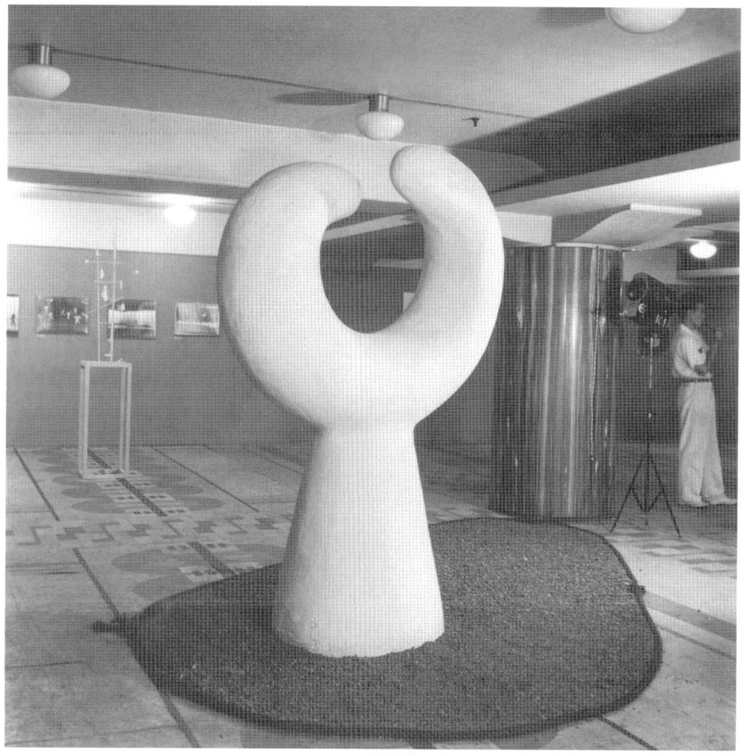

Isamu Noguchi's *Mu* at the exhibition of his ceramics and sculpture at Mitsukoshi Department Store, 1950.

Steps to the Bamboo Basket Chair

Why a bamboo basket chair in 1950? The IARI was established by the Ministry of Commerce and Industry in 1928 with two major missions: to promote the export of products made by traditional techniques, and to conduct scientific research in manufacturing technology. In 1933, when Bruno Taut was invited to teach the making of mass production prototypes, called master models, he emphasized two objectives: to study the structure and function of a shape, and to make an effort to extract modern form from Japanese tradition.[8] The weaving of plant fiber was a widely developed technique in Asia and Africa, with bamboo and rattan as the most common materials. Kenmochi, a worshipful student of Taut, was drawn to this conventional technique. Taut himself would later create his own designs using materials such as bamboo and *meisen* (silk cloth), after he relocated to Takasaki. Bamboo was not a precious material but rather a commonplace found in daily housewares, making it practical and worthwhile for the Institute to apply to modern designs and export goods.

From 1934 to 1938, Kenmochi was occupied with studies of weaving alongside the production of master models. Using bamboo and rattan, he completed Fruit Basket (1939) and Magazine Rack (1940). At first he straightforwardly applied conventional techniques to the making of modern shapes, but within a few years his techniques evolved to take advantage of the characteristics of the material in the structure of his products.

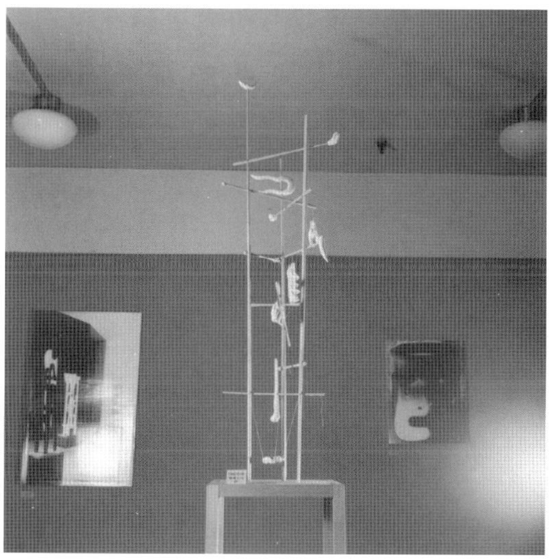

Isamu Noguchi, *Model of Bell Tower for Hiroshima*, Mitsukoshi Department Store exhibition, 1950.

From 1934, the pliable nature of bamboo was applied to experiments with cantilevers and curved surfaces.[9] The French designer Charlotte Perriand, another invitee to the Institute, also adopted weaving extensively. Perriand was captivated by bamboo weaving during her stay in Japan, and designed chairs and trays using this technique. A photograph of the Institute's prototype Bamboo Lounge Chair on a cover of *Industrial Art News* seems a clear reference to a similar shot of Perriand stretching over a bamboo lounge chair of her own design.

After the war, bamboo and rattan weaving were revived as part of Japanese economic reconstruction. The manufacturing industry had been damaged during the war, and in a country of poor resources it was an imperative challenge to make use of these techniques in creating new designs. The Institute also explored the possibility of using bamboo in composite

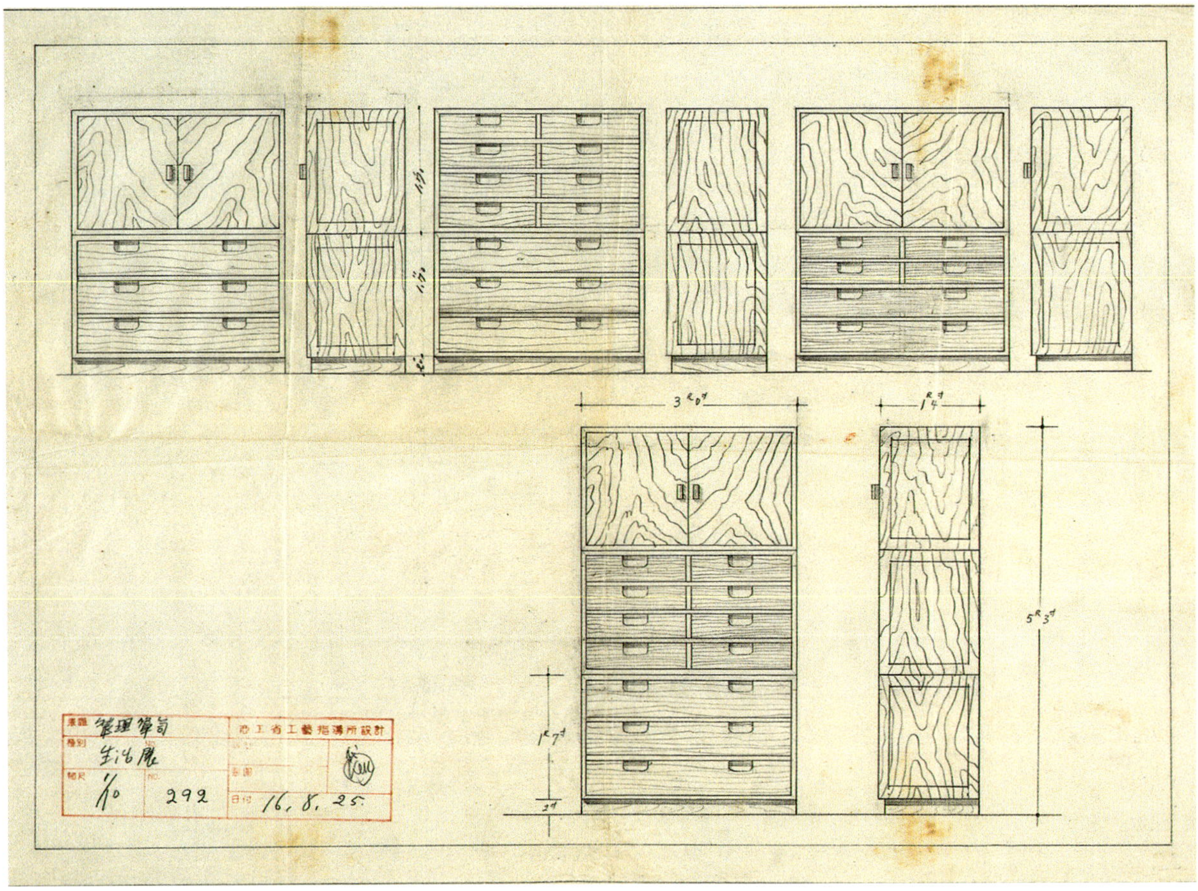

Isamu Kenmochi, Wardrobes, pencil on paper, 1941, 16 x 21 in.

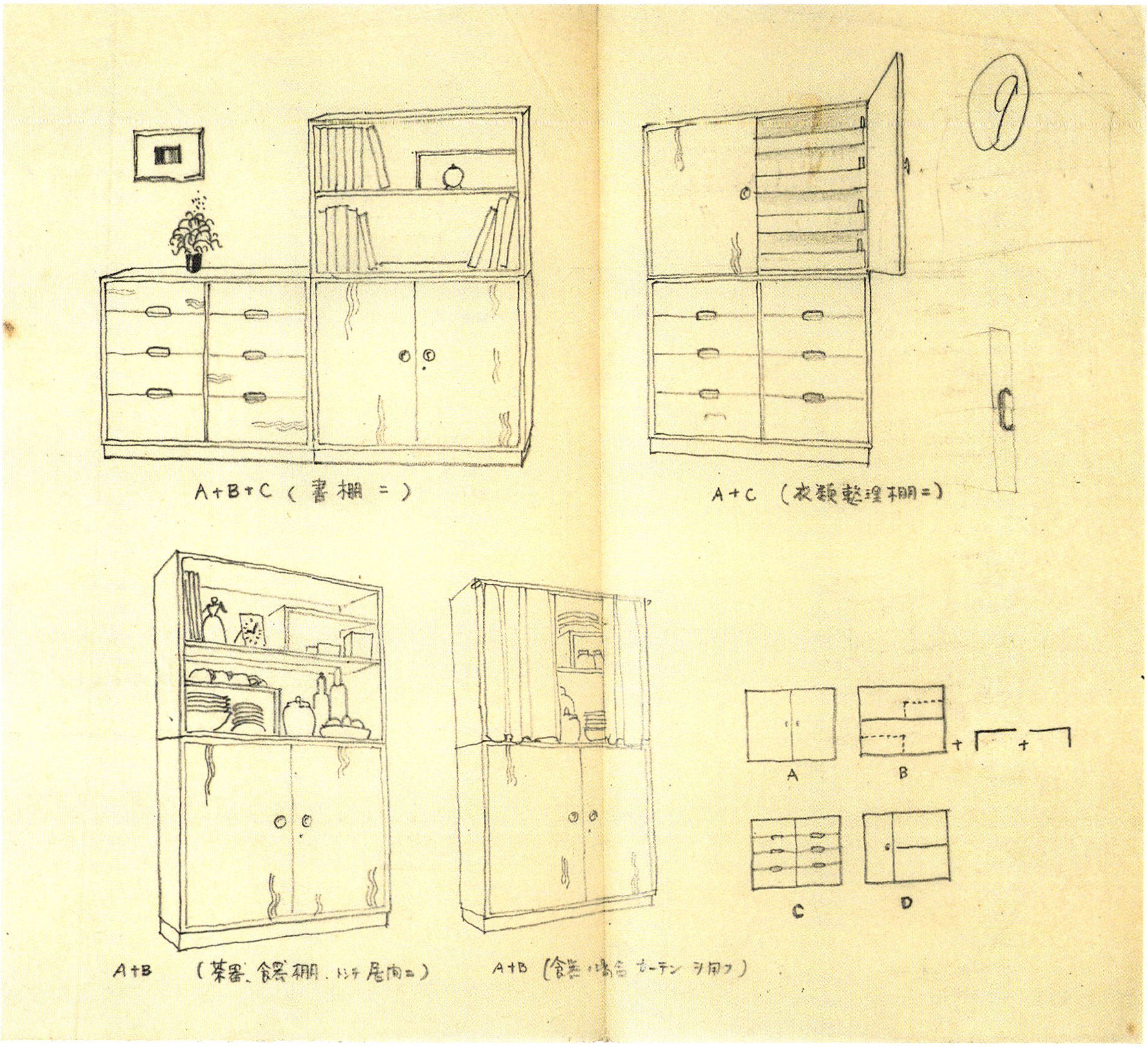

Isamu Kenmochi, Furniture Units, pencil on paper, ca. 1940s, 15 1/2 x 20 in. Kenmochi made a number of drawings of furniture while working at the Industrial Arts Research Institute. Founded in 1928, the Institute was the first government agency devoted to research and consultation in industrial design. Kenmochi's designs were based on the forms of traditional wooden Japanese wardrobes but were inspired by the flexibility and compatibility of modern furniture units.

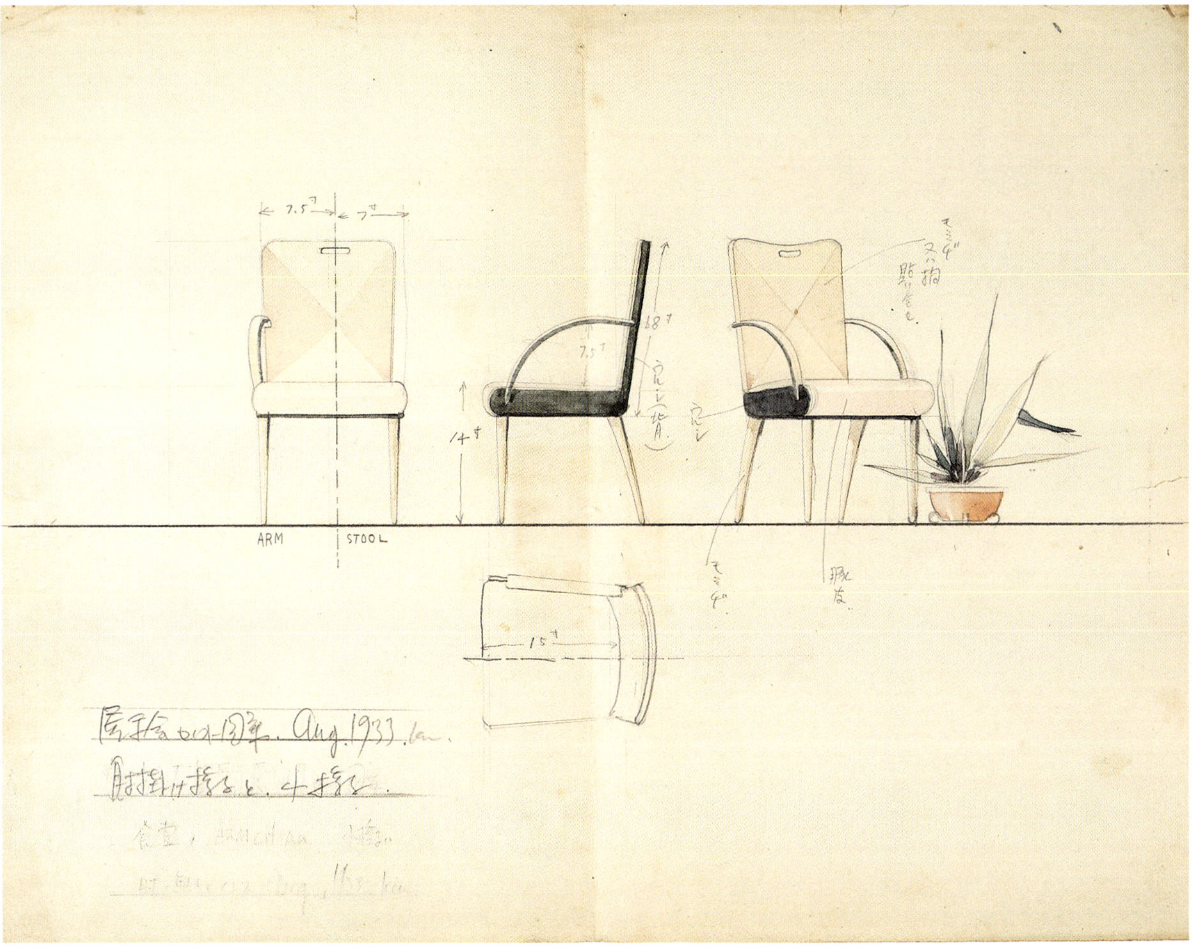

Isamu Kenmochi, Chair, pencil and gouache on paper, 1933, 12 x 13 in.

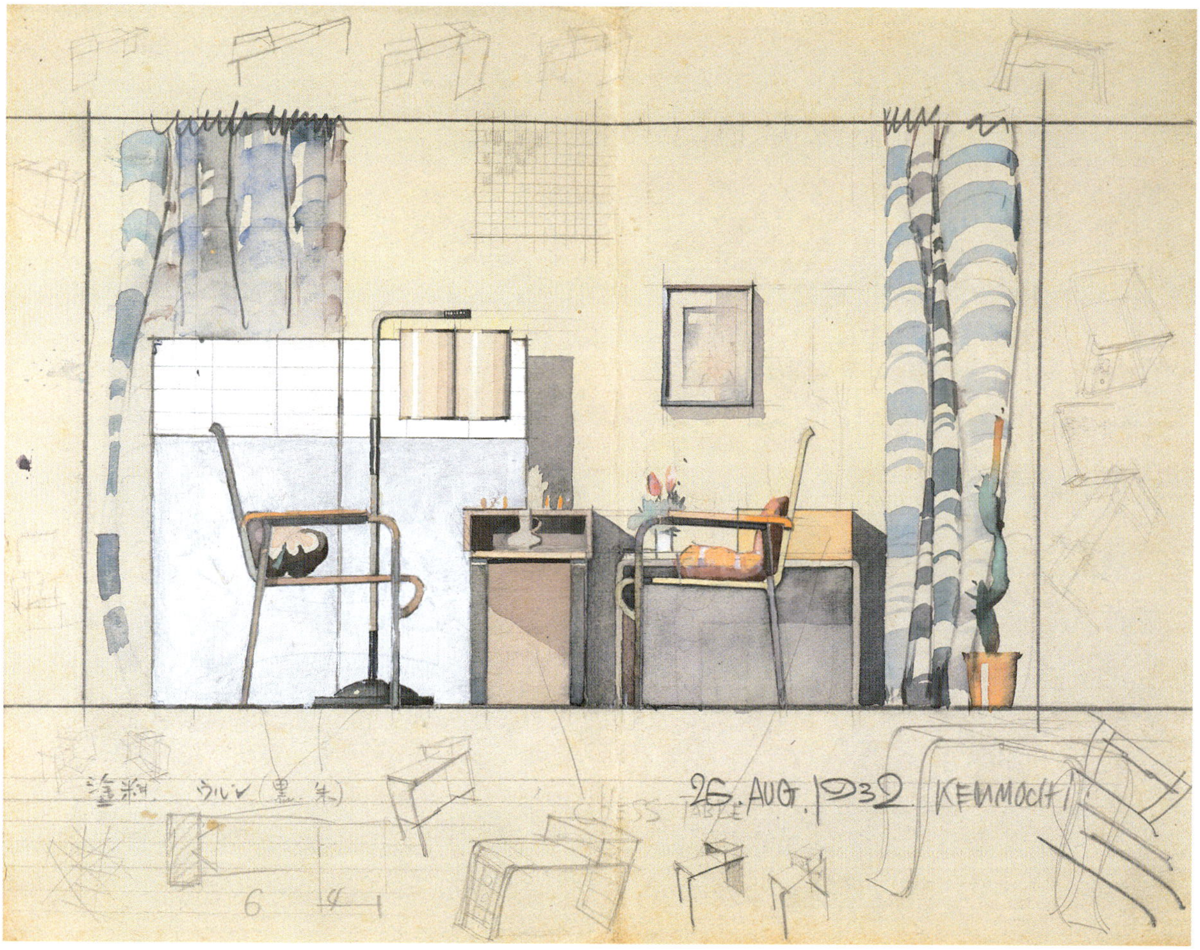

Isamu Kenmochi, Reception Room, pencil and gouache on paper, 1932, 12 x 14 in.

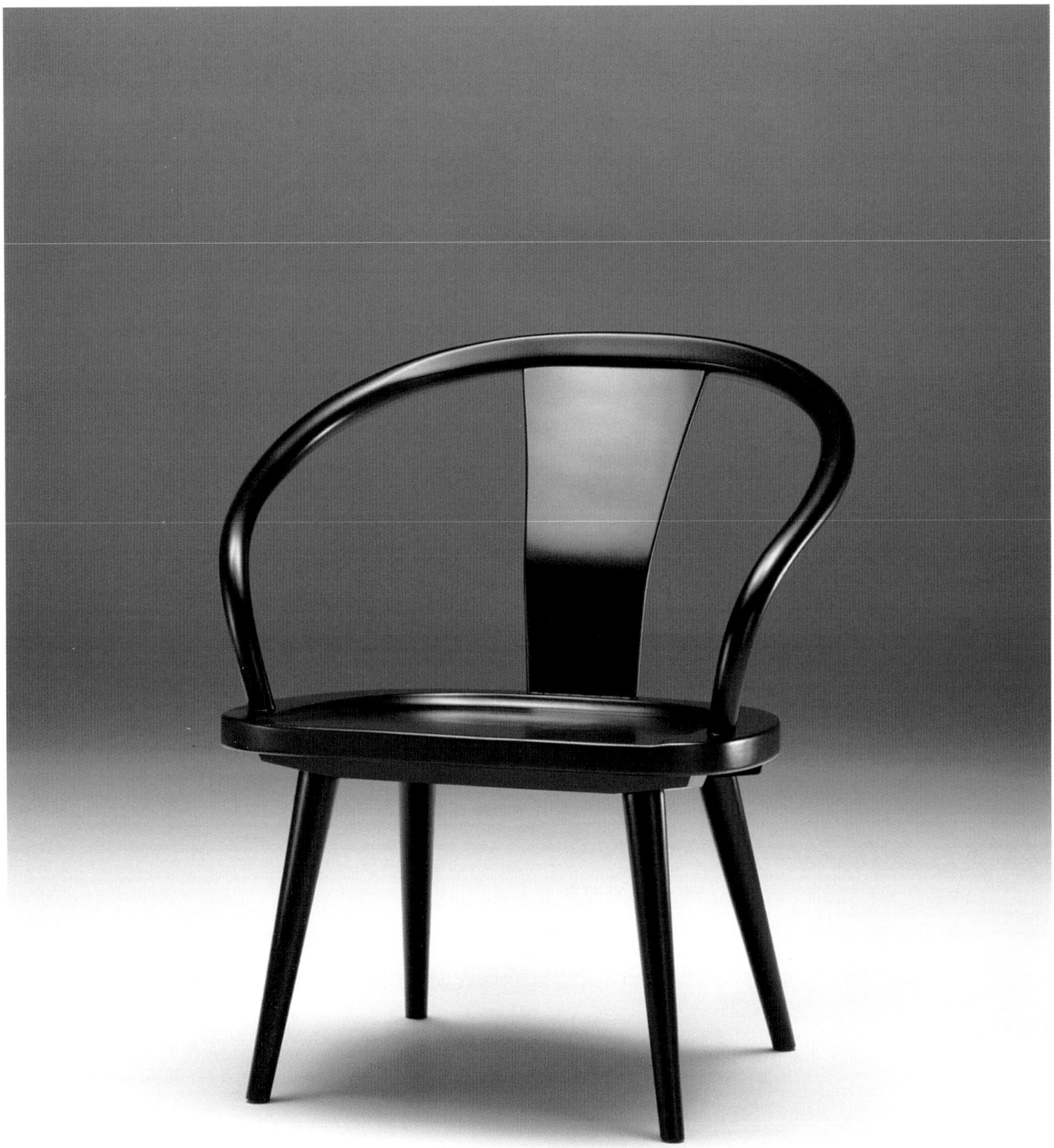

Kenmochi Design Associates, Armchair 207, beech wood, 1960, 27 15/16 x 22 13/16 x 24 15/16 in., manufactured by Akita Mokko Co., Ltd.

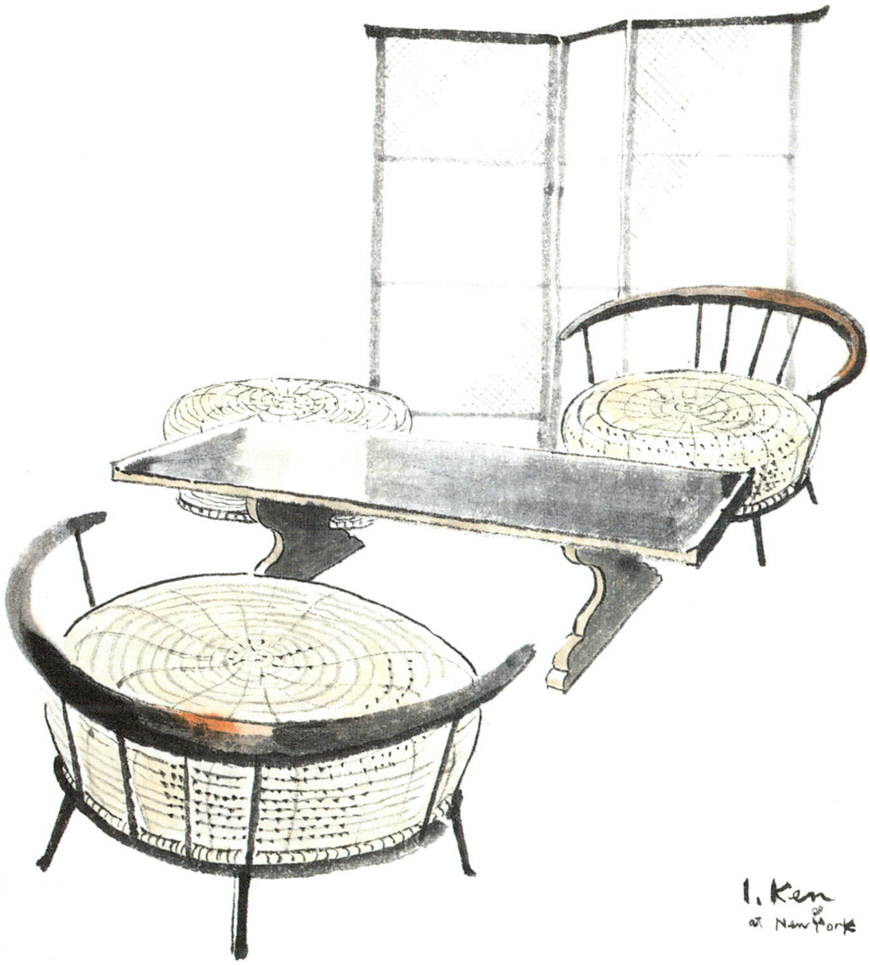

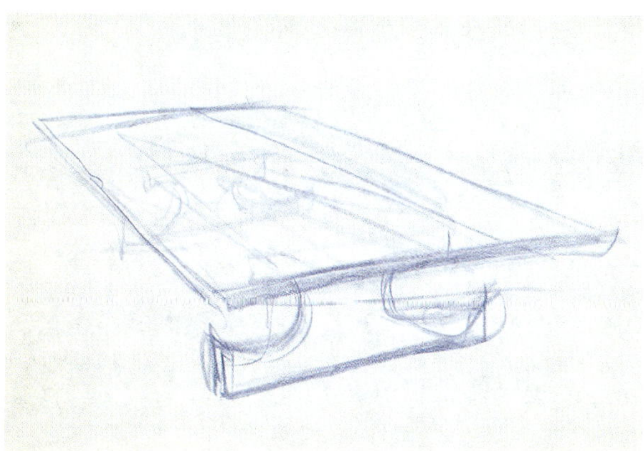

Top: Isamu Kenmochi, Shikisi, pencil on paper, 1953, 10 3/4 x 9 7/16 in. This drawing was made during Kenmochi's visit to the United States to research the latest trends in industrial design. Bottom: Isamu Kenmochi, Study Table, pencil on tracing paper, 1954, 8 x 9 1/2 in.

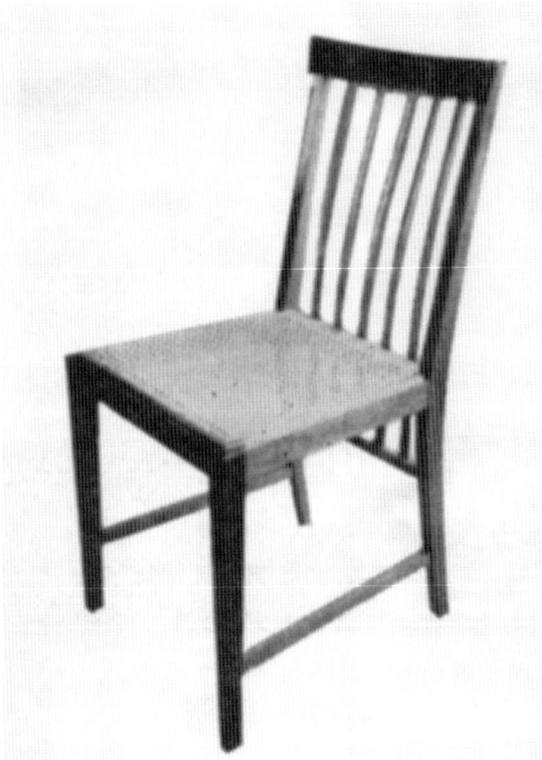

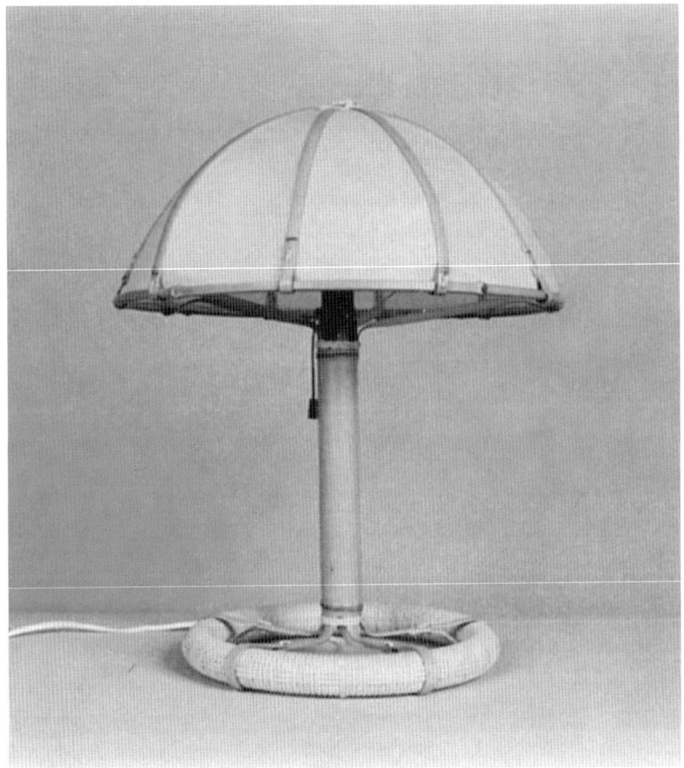

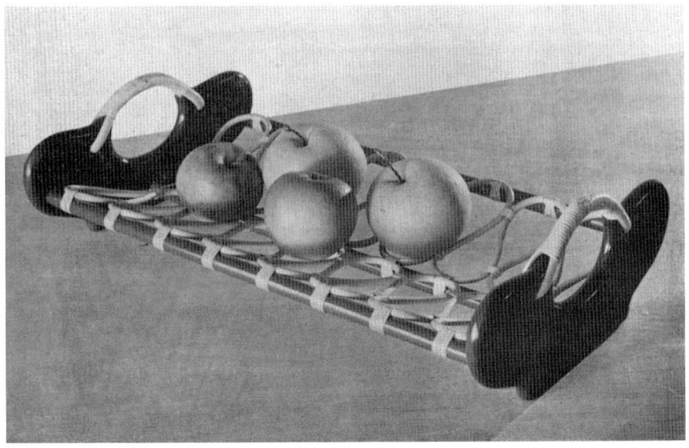

Clockwise from top left: Dining Chair Prototype made at Industrial Arts Research Institute, 1937; Bruno Taut, Bamboo Stand, 1934; Isamu Kenmochi, Fruit Basket, 1939.

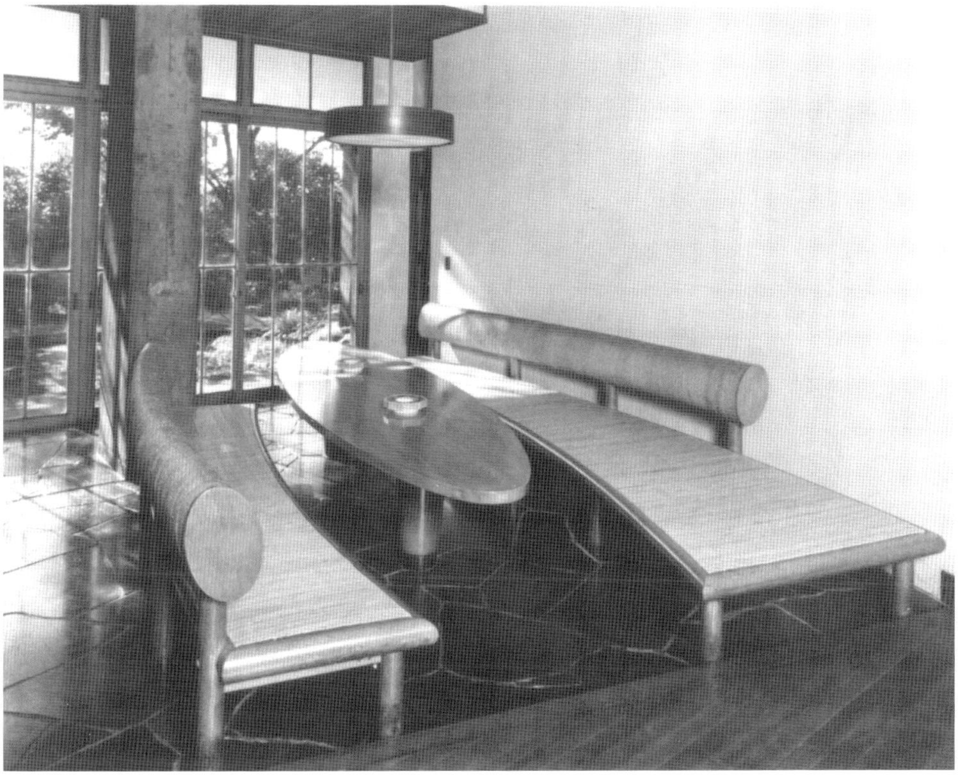

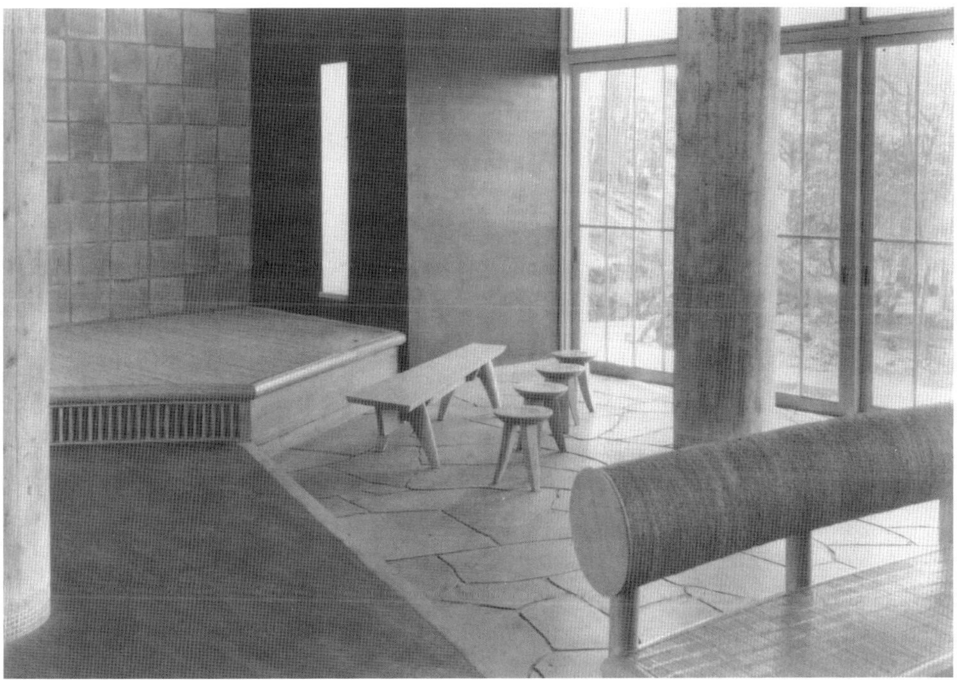

Top: Isamu Noguchi, Curved Tatami Bench, wood, tatami, rope, and copper, designed for the Faculty Room, *Shin Banraisha*, Keio University, 1950–1951. Bottom: Isamu Noguchi, Coffee Table and Three-Legged Stools, *Shin Banraisha*, 1950–1951.

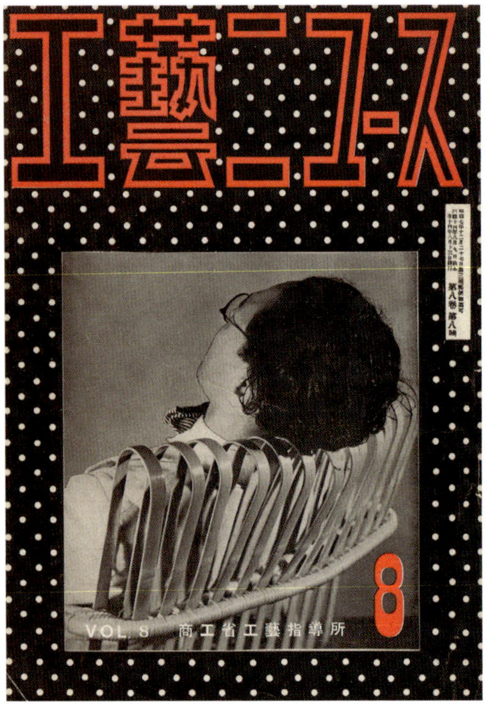

Bamboo Chair, cover of *Industrial Art News*, Vol. 8, No. 8, August, 1939.

materials. At the Institute's Kyushu branch, in Kurume City, Fukuoka Prefecture, in a major bamboo habitat region, a wide range of research was conducted as part of the main development project. In the realm of postwar furniture production, the curved surface was making its way through the new molded plywood technology, adding pressure to the development of traditional shapes and material elasticity. When the Japanese Bamboo Products exhibition opened in December 1949 at the British Council of Industrial Design in London, it won positive reviews, to the encouragement of designers in Japan.[10]

Beyond the "Tradition Controversy"

Through these developmental stages, we are able to understand the eagerness and expectation Kenmochi had when he spoke to Noguchi about bamboo. The fact that their Bamboo Basket Chair was released under Noguchi's name provoked the "Tradition Controversy," which would be actively debated throughout the 1950s. First, there was Noguchi, who came to his fatherland as the "most avant-garde artist from America," the most influential country in postwar Japan, and expressed his affection for Japanese culture and reevaluated it. I will not go into the matter of "Tradition Controversy" here. But in the 1950s, Japan was also at the frontline of the tension and conflict between East and West. Internally, the progressives, newly freed from government pressure, regained strength to cooperate with the Socialists and were eager to reconstruct Japanese culture as part of a new "Nationalistic Culture," all in the whirlwind of the colonial independence movement. More accurately, there were some Internationalists who could not be completely anti-establishment during the war, and still admired Western universality as progress. Having learned their lesson, they could not deny that Japan's modernization would proceed on its own. Then in 1953, when Ichiro Hariu translated Jôzsef Révai's criticism of Hungarian functionalism as "worship for decadent machinery," Japanese architects were swayed significantly by it.[11] In hindsight, Révai was little more than a soft Stalinist who embraced the idea that "architecture in the Soviet Union must serve new Socialists," but at the time the progressives were either prewar Marxists or Internationalists sympathetic to the Marxists—in other words, believers in International Architecture—and they were in no position to reject Révai's assertion.

New Japanese Style (*Kindai Nihon cho*), proposed in 1952 by Kenmochi upon his return from the United States, became a target of criticism as Japonica by both Socialists and Internationalists alike, for it was neither an advocacy of the internationalization of Japanese culture nor a criticism of Japanese culture. It was an act of amalgamating tradition with the latest modern designs, skipping Japanese tradition and criticism. It was a departure from the nineteenth-century outlook on the world. In the background, there was the deep confrontational structure of East and West, which had taken over the postwar politics. The effort of these two

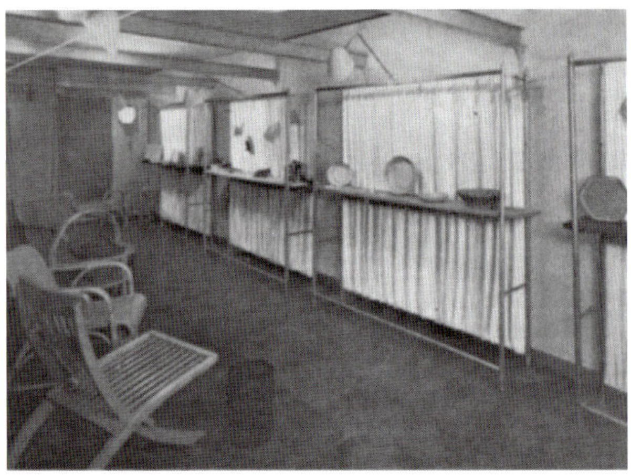

Installation view, Japanese Bamboo Products exhibition, London, December 1949.

men, who were neither Socialists nor Ethnicists, was analogous to building a bridge over this crevasse.

On one hand, Toshiro Mayuzumi's ethnistic proposal to reject the "easy national identity" of the Soviets and to "focus on the creative emotions of oneself as Japanese" matched Kenmochi's position in direction and intent.[12] Yet around this time, the sensibility expressed by Noguchi through his Akari series (1951–1986) was not a simple revival of tradition, but the sublimation of traditional materials into modern design. This, as with the work of Perriand, was regarded as a foreigner's discovery of tradition and did not bring about the same criticism as Kenmochi's elaboration of Japanese Modern. Still, the works of Noguchi and Kenmochi, in their attempt to modernize tradition, remain in the same trajectory that originated at the time of their collaborative effort. Their chance meeting may have been inevitable, but the fact that the foundation and conception of their creativity were so completely aligned, as if by a greater will, is nothing short of a miracle. This, I think, is what gives their ambitions and forms a timeless tenure, alive today and for the future.

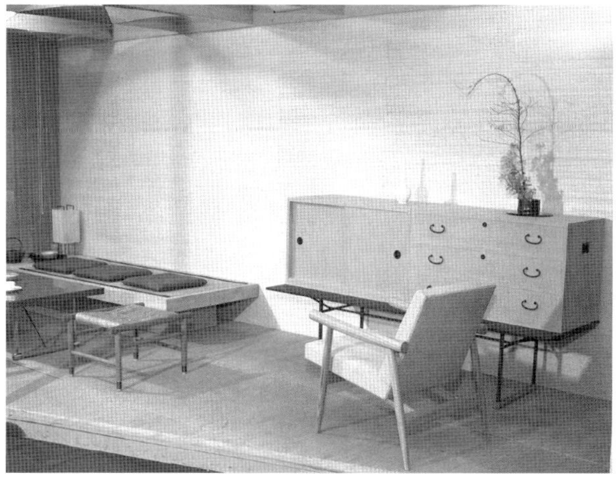

Isamu Kenmochi, Model Room, 1954.

Translated from the Japanese by Eiko Sakai and Wayne and Noriko Peet.

NOTES

1. Isamu Kenmochi, "Isamu Noguchi as I See Him," *Industrial Art News,* vol. 18, no. 10 (October 1950): 22.
2. Isamu Noguchi, "Art and Community," *Bijutsu Techo* (July 1950).
3. Saburo Hasegawa, "Meeting Isamu Noguchi," *Bijutsu Techo* (July 1950).
4. Genichiro Inokuma, "Mu," *Mizuwe*, no. 540 (October 1950).
5. Kenmochi, *Industrial Art News,* vol. 18, no. 10 (October 1950): 22.
6. It wasn't until the late 1950s that molded plywood technology could be applied to production in Japan.
7. Kenmochi, *Industrial Art News,* vol. 18, no. 10 (October 1950): 22.
8. Hitoshi Mori, "Functionalist Design in Wartime," from 2003–2004 Grant-in-Aid for Scientific Research B (1), "Report: The War, Symbol, and Art in the 20th Century through Exhibitions, Films, Prints, and Products," Chiba University, 2005.
9. "Bamboo Furniture in Japan from 1932 to the Present," *Industrial Art News*, vol 18, no. 7 (July 1950).
10. Shinji Koike, "Japanese Bamboo Products Exhibited in London," *Industrial Art News*, vol. 18, no. 7 (July 1950). In the same year, Junzo Sakakura announced his bamboo chair almost identical to that of Perriand's. It is of interest that Sakakura, an Internationalist, was attracted to the same trend as Kenmochi at the same time, and to see the standard he sought.
11. Jôzsef Révai, "The Traditions of Architecture and Modernism," *Bijutsu Hihyo* (October 1953).
12. Toshiro Mayuzumi, "Art to Be Sat On," *Bijutsu Hihyo* (May 1955).

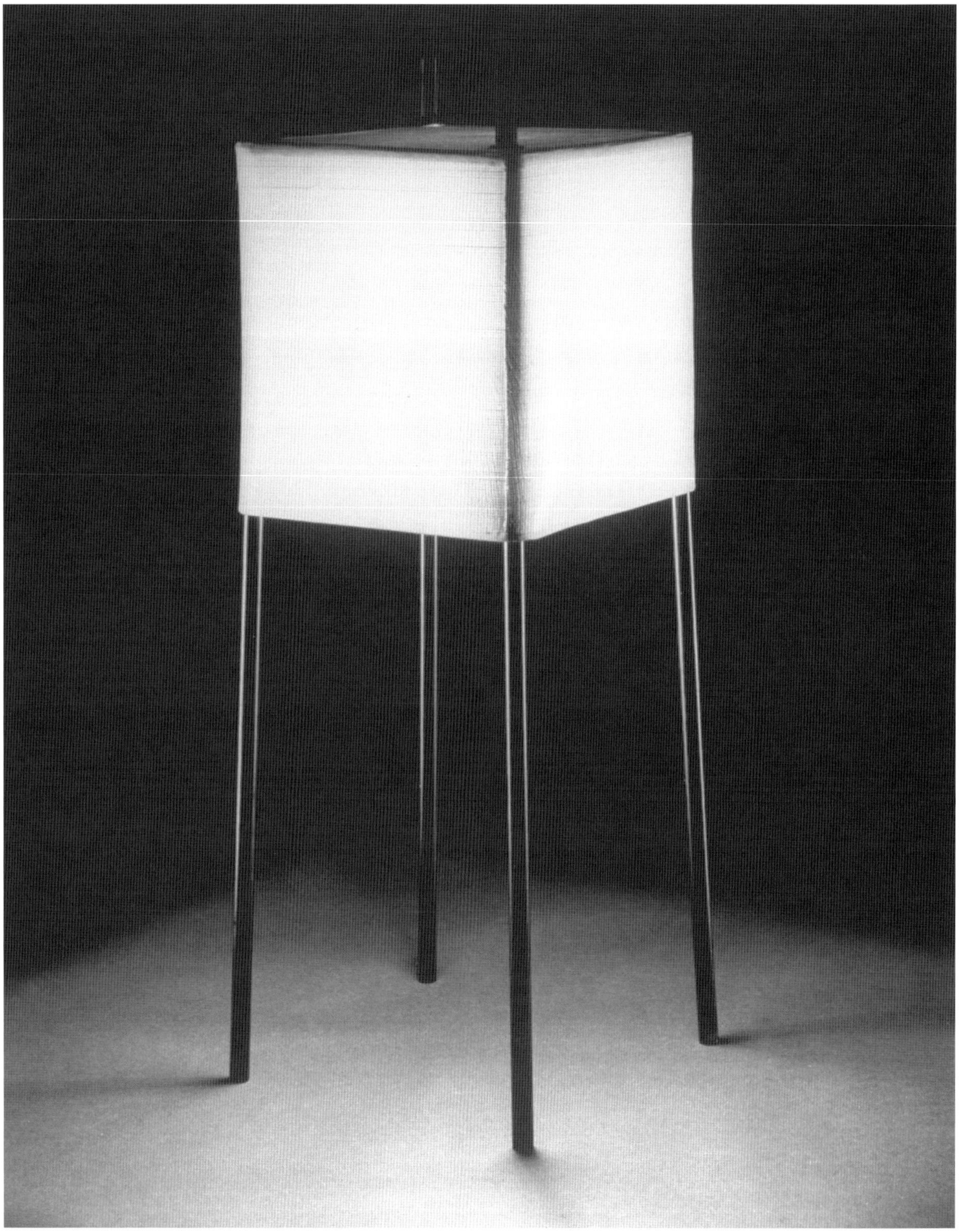

Isamu Noguchi, Four-Legged Lamp, oiled parchment shade and iron, ca. 1945, height: 30 in. Marketed by the company Lightolier as a low-priced readymade, this lamp created no glare when light showed through the parchment.

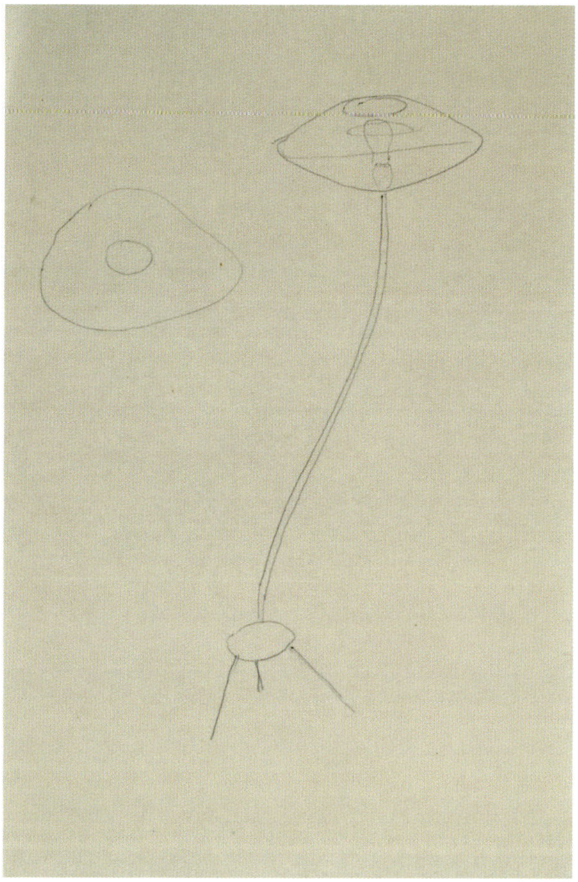

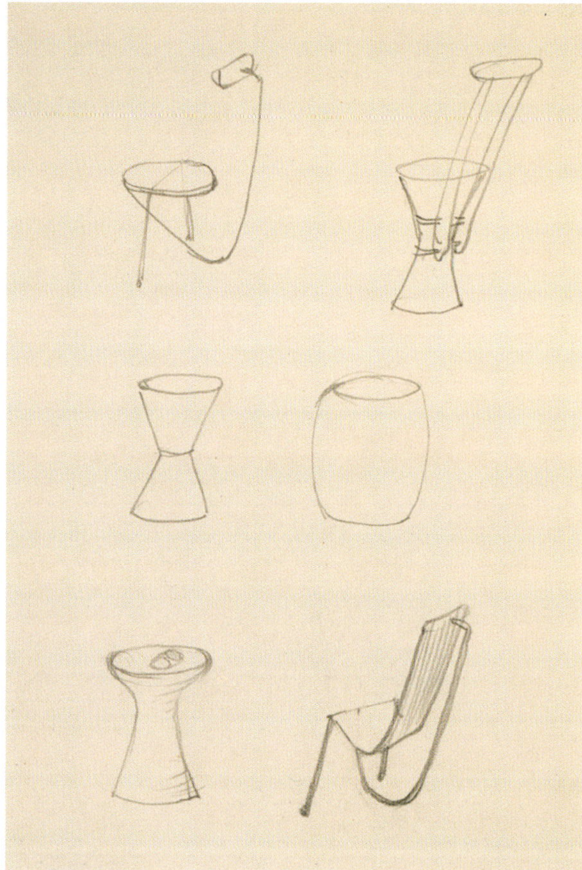

Above left: Isamu Noguchi, Lamp, ink on paper, ca. 1942–1948, 8 1/2 x 11 in. Above right: Isamu Noguchi, Six Ideas for Sculpture and Furniture, pencil on paper, ca. 1942–1948, 8 x 4 15/16 in.

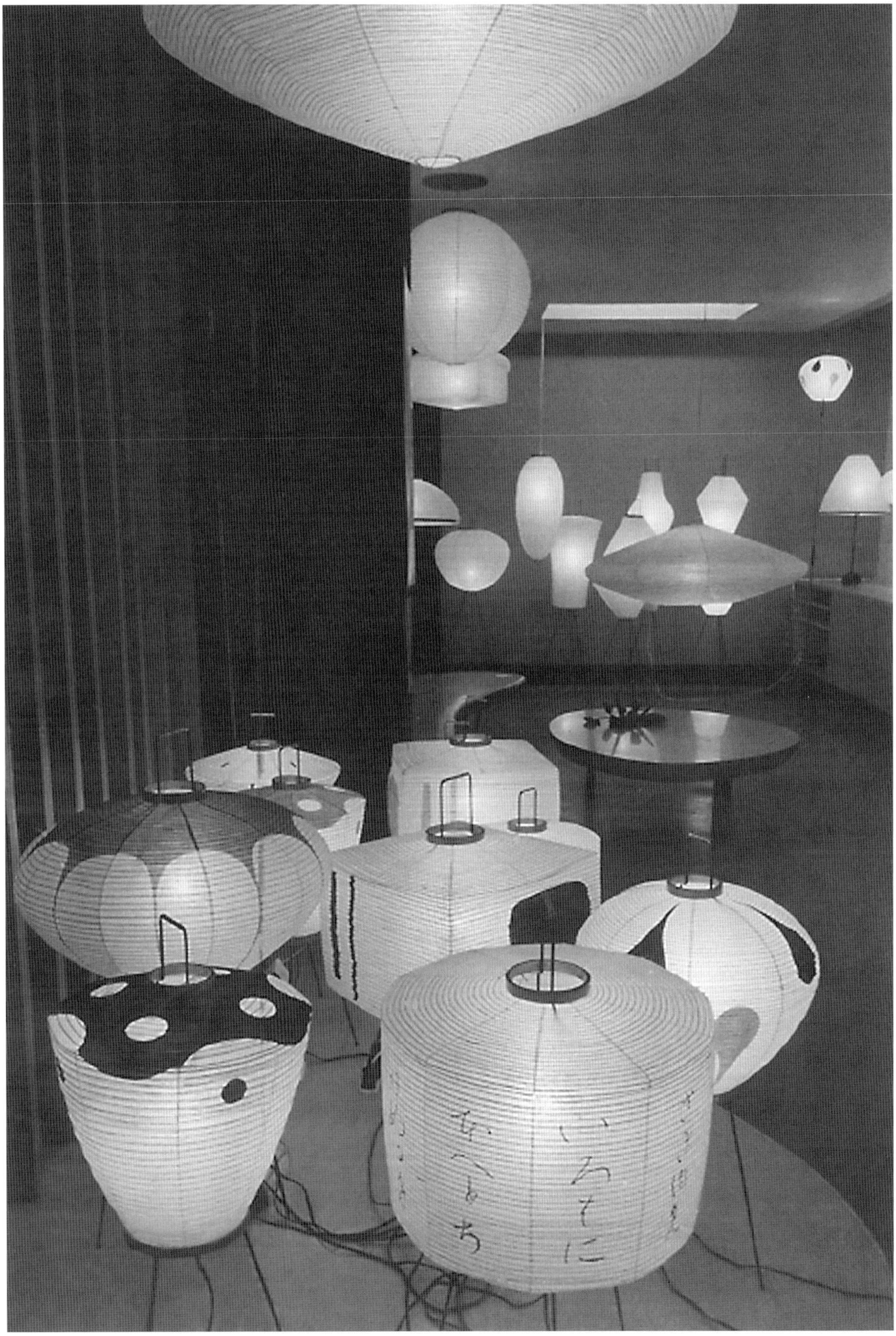

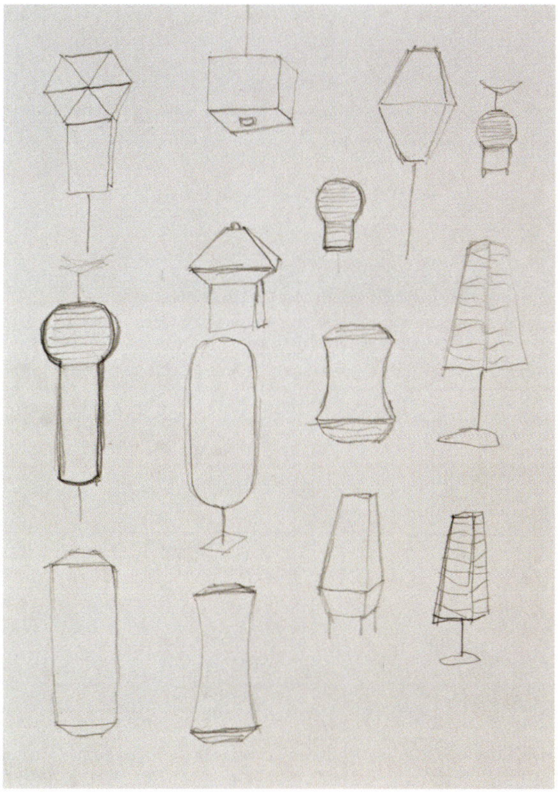

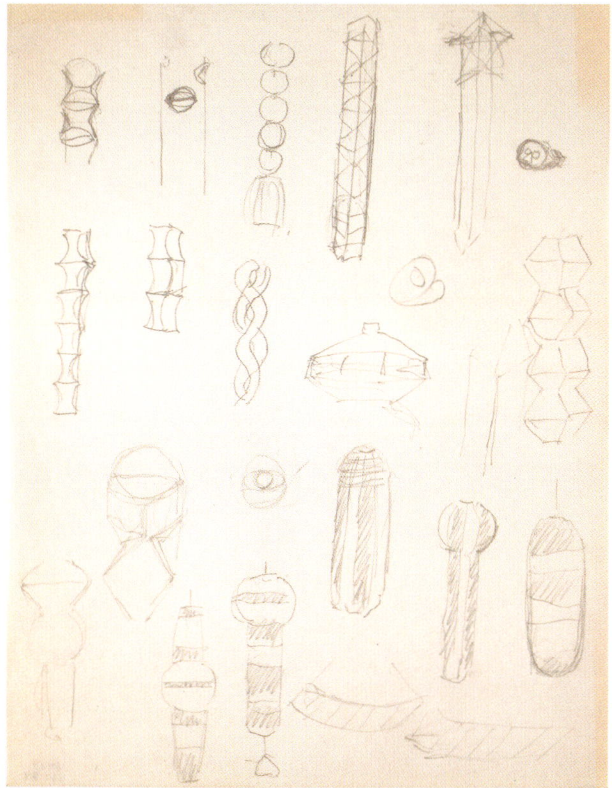

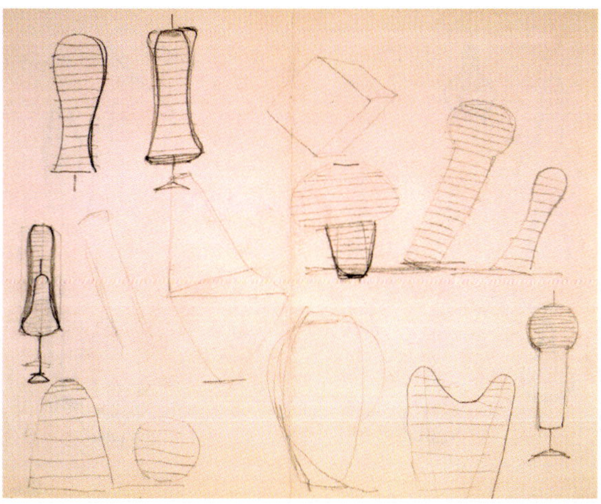

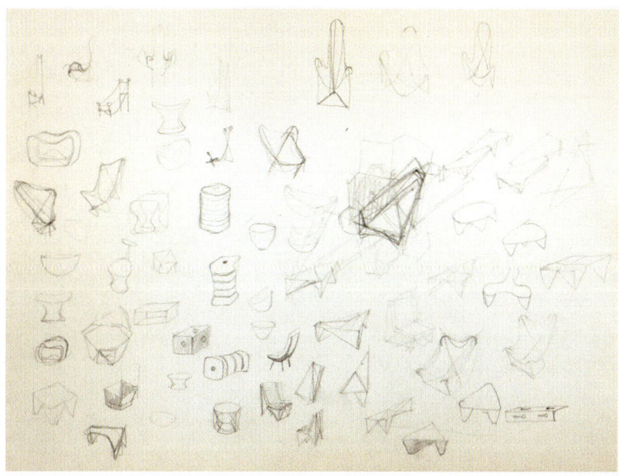

Clockwise from top left: Isamu Noguchi, Akari Drawing, pencil on paper, ca. 1965–1969, 10 11/16 x 8 3/16 in.; Ideas for Sculpture and Akari, pencil on paper, ca. 1965, 11 x 14 in.; Furniture Drawing from a Sketchbook, graphite on paper, 1952, 19 3/4 x 14 3/4 in.; Akari Drawing, pencil on paper, 1960, 11 x 13 5/8 in. Opposite: Isamu Noguchi, early Akari Light Sculptures, shown at Chuo Koron Gallery, 1954.

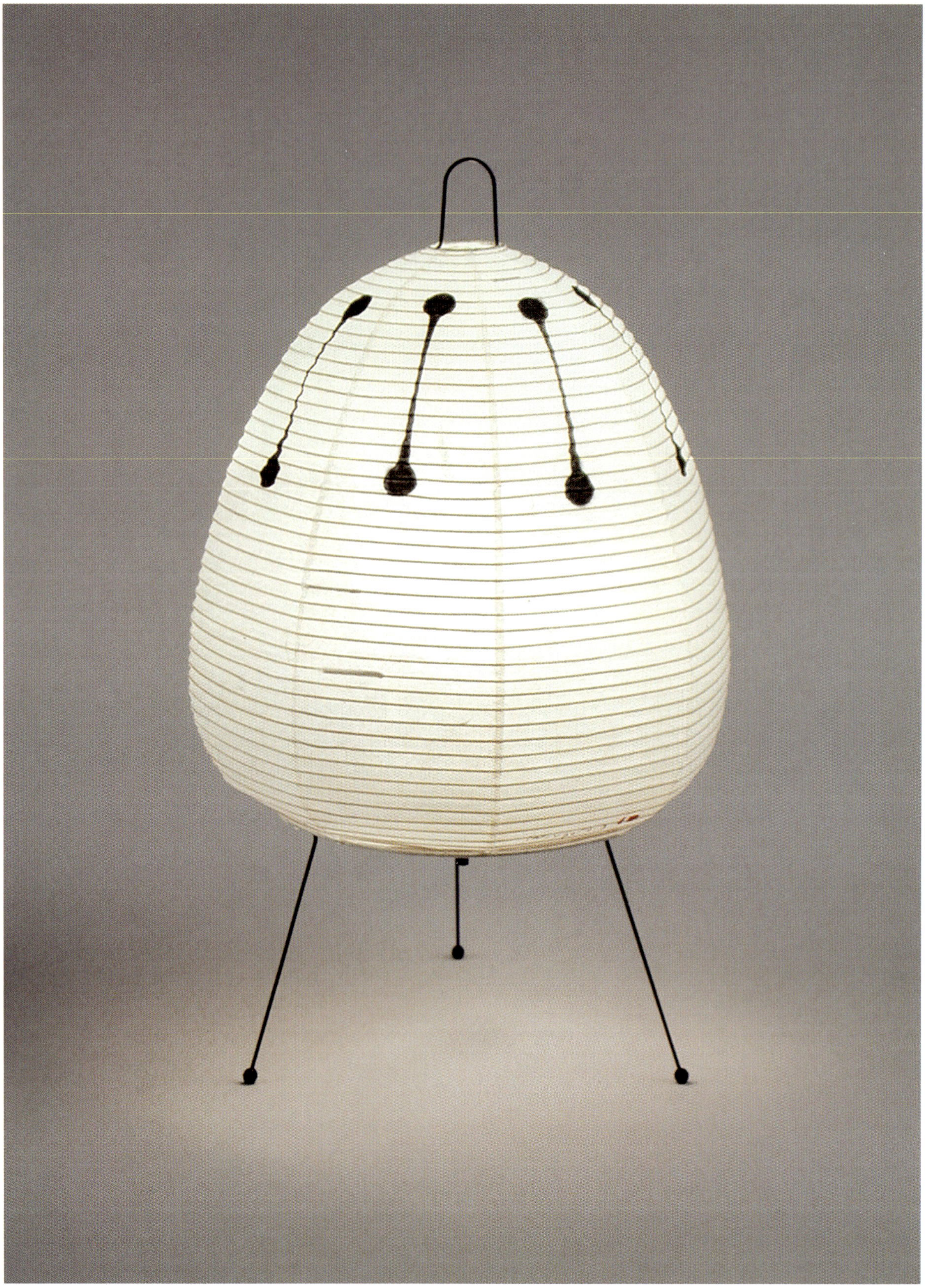

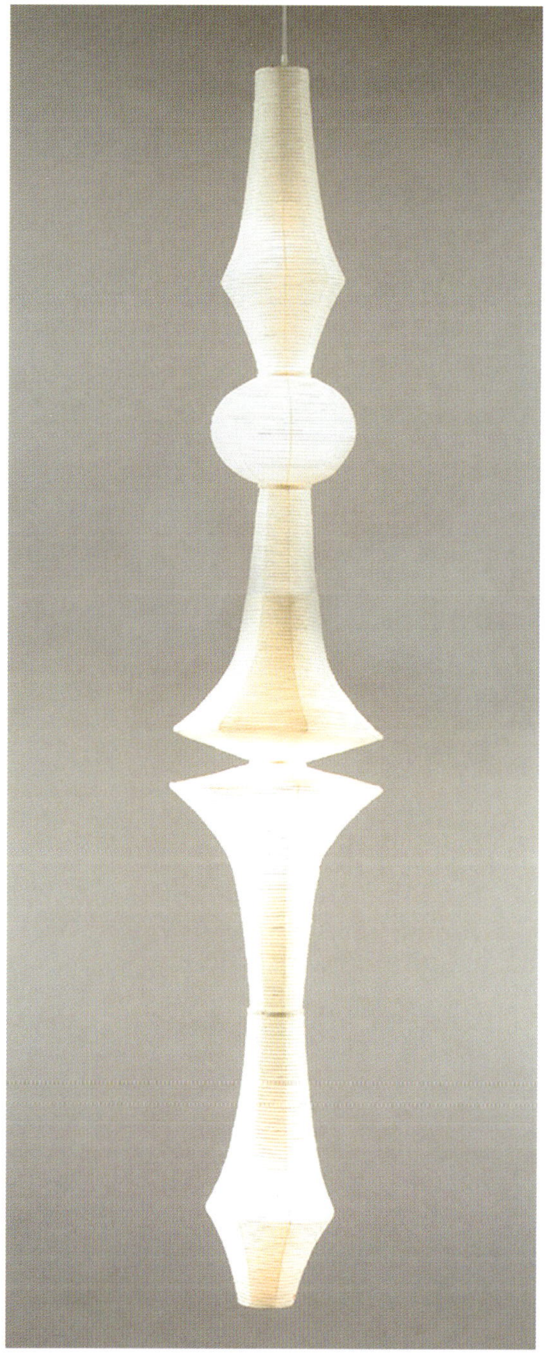

Above: Isamu Noguchi, Akari E, Washi paper and bamboo ribbing, 1954, height: 116 in., diameter: 19 in., manufactured by Ozeki & Co., Ltd. Opposite: Akari 1AD, Washi paper, bamboo ribbing, and metal, 1952, height: 17 in., diameter: 10 in., manufactured by Ozeki & Co., Ltd. Noguchi's early Akari were produced in Gifu, Japan, by the Ozeki family, whose firm had manufactured paper lanterns since the Meiji period and continues to produce Akari today. The lamps were first exhibited at the Kamakura Museum of Modern Art in 1952, and later at the Chuo Koron Gallery in Tokyo in 1954, which also featured an interior designed by Noguchi. Following the success of the Kamakura exhibition, Noguchi worked to arrange international distribution of Akari. He spoke first with Lightolier, but in 1953 signed with Bonniers on Madison Avenue in New York, which held the first international exhibition of Akari in spring 1955.

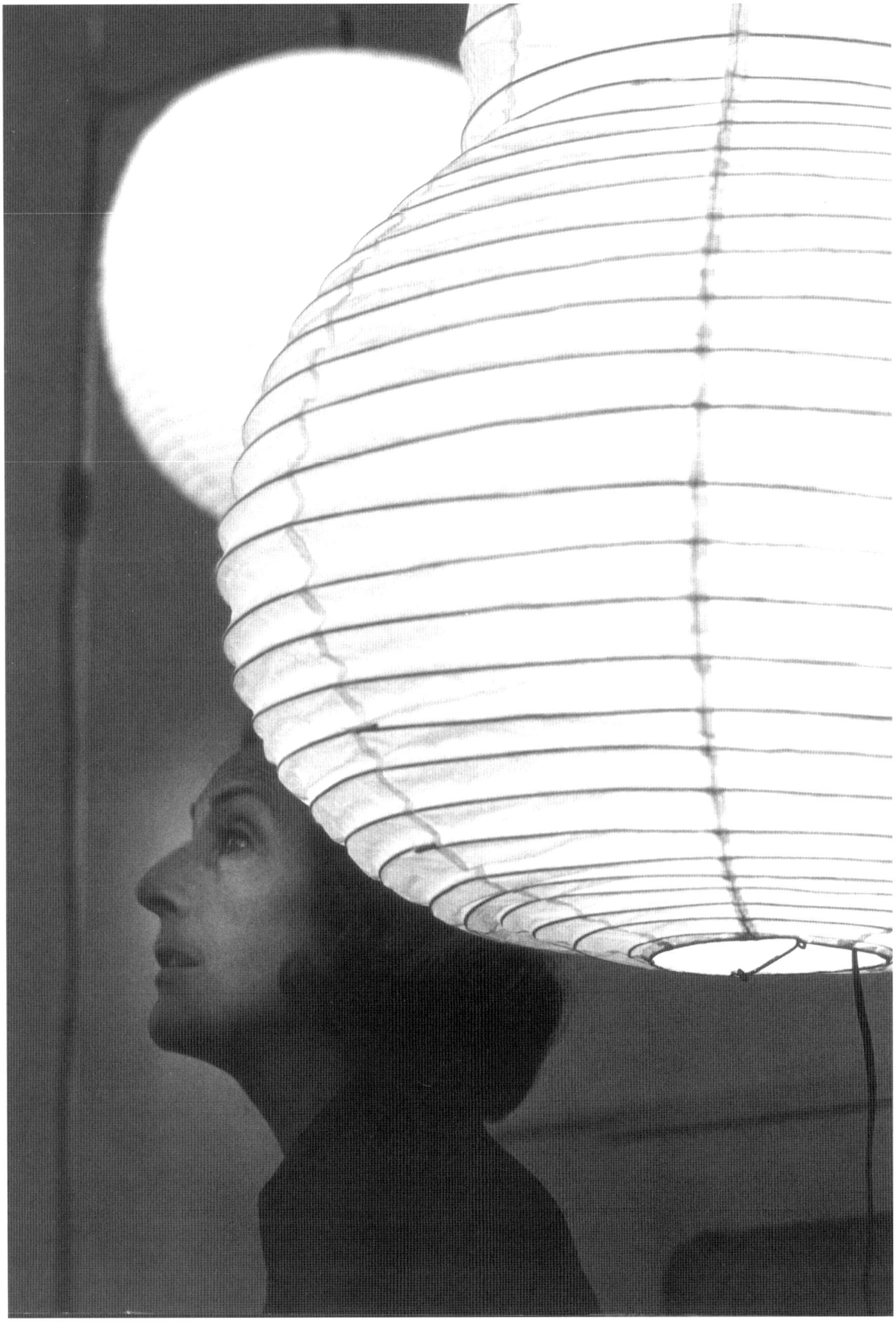

Woman looking at Akari, Cordier and Ekstrom Gallery, New York, 1968.

TOWARD (RE)DEFINING AKARI

NINA MURAYAMA

MY THEORY ON TRADITION IS THAT IT CAN ALWAYS CHANGE. Isamu Noguchi, *New York Times*, April 6, 1970.

I AM A PRACTICAL ARTIST. Isamu Noguchi, *Art in America*, March-April, 1968.

Isamu Noguchi had a long career as a sculptor, stage set designer, industrial designer, and landscape designer. Fundamentally, his involvement with design was motivated by his aesthetic concern for spaces, and his belief in the artist's role of serving society. Noguchi's consciousness about the way aspects of his cultural identity emerged in his work developed as he traveled to Japan in 1931 and 1950, interacted with artists, designers, and architects, and was inspired by the cultural heritage of Japan. Noguchi had a propensity for traveling throughout his life; he perhaps felt at home everywhere because he belonged nowhere. Likewise, Noguchi's place in modern design and art discourse keeps shifting between categories: Brancusian functional sculpture, Surrealism, Good Design, or Japanese Modern. Exploring borderless artistic exchanges and experiments at various sites such as New York, Mure, Kyoto, Paris, Querceta, Italy, Beijing, and New Mexico, Noguchi trained himself to penetrate the essential nature of the materials, forms, techniques, and structures used in artworks and functional objects from various regions. Together with Isamu Kenmochi, Noguchi avoided Japonica, defined broadly as a schematic adaptation of formal elements found in traditional Japanese everyday objects and architecture that appealed mainly to eyes looking for exotic foreign features. Their collaboration in the summer of 1950 confirmed Noguchi's cross-cultural design principle of working locally with a global perspective to negotiate cultural differences.

Noguchi was destined to lead a borderless way of life, as reflected in his creations. Noguchi spent more than ten years of his childhood in Japan with his mother, Leonie Gilmour, an American writer, and mostly they lived separately from his father, Yonejiro Noguchi, a Japanese poet. In 1918, at the age of thirteen, he was sent back alone to the United States from Japan by his mother. After Noguchi left Japan, the country increasingly became a foreign and remote place for him, especially after he felt betrayed by his father during his 1931 visit, when he faced Yonejiro's reluctance to accept his half-American son who had been born in the United States. Long after this bitter trip, Noguchi revisited Japan in 1950 with mixed feelings fueled by fear and an eagerness to reconnect to the country. His father had already died, and Japan was struggling to rise up from the rubble of World War II. This visit opened up a new relationship for Noguchi with the country, and also marked the beginning of his most productive decade, in which he generated a variety of furniture and other design products, including his first design of a Japanese lantern, which he called *akari*, the Japanese word for light.

Noguchi's approach to design was radical in that he tried to redefine a concept for each functional object or architectural element spatially and conceptually. His design principle was outlined in the following statement he addressed to Japanese audiences in 1950: "A window need not be merely a window. If you think of a window as a wall of light, then you can eliminate the conventional idea of a window and now it becomes a wall that is also a source of light."[1] In other words, the concept of a window is not merely about its shape or style, but about an architectural element linking inside and outside spaces. Furthermore, Noguchi was aware of the originality

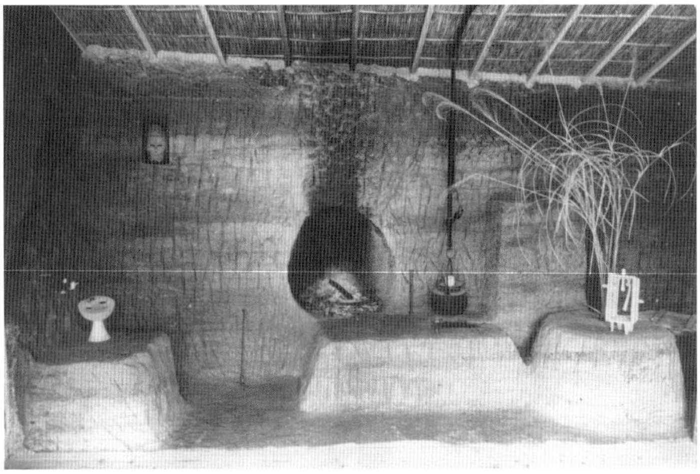

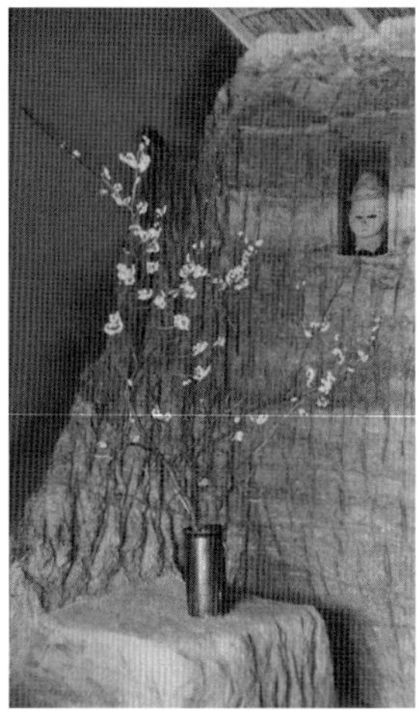

Above left: View of Noguchi's studio at Kita Kamakura, Japan, 1952. Above right: Inspired by the Japanese architectural feature called *tokonoma*, Noguchi redesigned his studio wall with a niche to display artworks.

of the Japanese cultural heritage and believed in its contribution to the development of postwar modern design. He was critical of the rapid modernization of new Tokyo while paying attention to Japanese cultural traditions discovered in Kyoto. In a letter dated May 16, 1950, he wrote about his feelings for the new and old Japan to his friend and accountant Bernard Bergstein:

> I have never had a reception such as I have received in Japan, and am pulled this way and that by all sorts of people and groups. They treat me like some sort of banner or symbol of their possible renaissance. There is a desperate hunger here for some sort of contact with the outside world, which is so touching that one does everything possible to help. The result of all these parties and speeches has been rather disastrous for my health which gets worse rather than better. This is one good reason that I am running off to Kyoto the day after tomorrow. That old capital I hear was not bombed and remains much as before the old Japan not this desperate new Japan of Tokyo.[2]

On his trip to view old temples and cultural treasures in Kyoto, Noguchi took pictures of a priest's chair. The way Noguchi captured the chair illustrates how he observed this simple yet dignified and sophisticated form, and shows his enthusiasm for its structure and function. This enthusiasm was typical of his attraction to the seemingly nondescript yet unforgettable formal features of everyday objects. Noguchi and Kenmochi shared a love and admiration for crafts and anonymous objects found in local communities. Noguchi visited Kenmochi at the Industrial Arts Research Institute in Tokyo on July 28, 1950, and spent two weeks that summer at the Institute's workshop, designing a prototype Bamboo Basket Chair based on Kenmochi's initial experiment with a bamboo weaving technique for

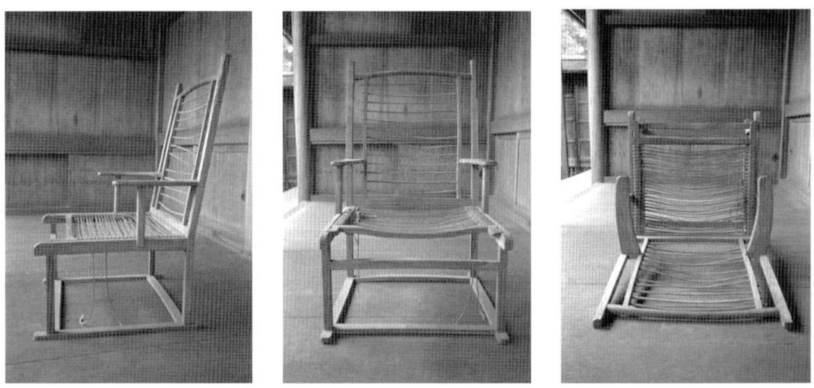

Photographs: Isamu Noguchi, a chair at a temple in Kyoto, Japan, ca. 1950.

the seat and back of the chair. Noguchi also made furniture for *Shin Banraisha*, as well as sculptures, models, and other works. Having closely witnessed Noguchi's absorption into creating many objects in a short period of time, Kenmochi felt that Noguchi's working attitude was akin to the rigorous devotion common in spiritual practices in the East. To a degree, Kenmochi even had the illusion that Noguchi was more of an Asian, while he and the other engineers at the Institute, promoting a new modern living environment through their design of tables and chairs, were more like Westerners.[3] It is true that Noguchi, through his experiences in Japan, avidly championed a Japanese cultural identity as expressed in modern design, as he wrote in the November 1950, issue of *Art and Architecture* after he returned to New York:

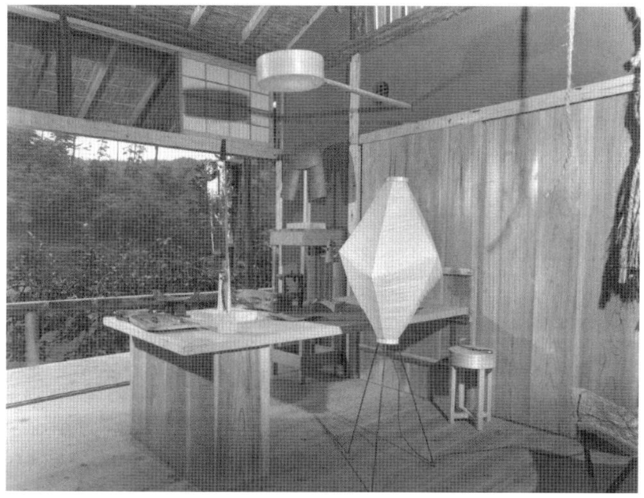

Noguchi's studio in Kita Kamakura, Japan, 1952.

> I used whatever materials were at hand, mainly the earth itself, or bamboo, or wood, or plaster to show the proportions of an object to be cut in stone. Indeed, when all the possibilities of modern technologies are lost, one returns once more to basic things, to basic materials, to basic thoughts. One starts all over again, and I think it [is] good. A people who have lost so much may gain this very great truth if they have the strength. . . . I suggested that to be modern did not mean to copy us but to be themselves, looking to their own roots for strength and inspiration.[4]

Although Kenmochi thought of Noguchi as being more Japanese than the Japanese themselves, in reality the works Noguchi created in Japan were mostly received as difficult to interpret in the context of either the Japan or the United States. Responding to Noguchi's show at the Museum of Modern Art in Kamakura, Shuzo Takiguchi, an art critic, pointed out in 1953, "From an American viewpoint, [Noguchi] is probably regarded as an artist of Oriental nature, but from our judgment of his works we detect a considerable amount of Western elements foreign to us."[5] Moreover, the historical context was ironic in that at the time Noguchi was advising Japanese artists to create works in harmony with their living environment, the postwar advanced art trend in Japan was leading toward bigger canvases aimed at display in museum spaces rather than in the common domestic sphere. Meanwhile, although Noguchi deeply respected elements of Japanese cultural heritage, such as gardens, *haniwa*, *ikebana*, tatami, and *tokonoma*, he couldn't deny his half-American blood, which made it difficult for him to identify completely with a Japanese cultural identity. He wrote in a letter to Bernard Bergstein dated April 24, 1951:

> I write this while waiting for a friend at airport, which explains the stationary [sic]. Seeing the planes coming and going I get a sense of spaces and vistas which one looses [sic] track of here in Japan where everything is so small and in Tokyo so drab. Indeed I feel homesick for the States, not N.Y. so much as the vast open spaces of the West.[6]

Noguchi had ambivalent feelings about his shifting cultural identification with Japan and the United States, but his cross-cultural working habits inspired his continued reinvention of traditional objects, mediums, and techniques in the works he made in different locations. Curiously, in his essay on Japanese Modern design, Kenmochi recognized Noguchi's Akari Light Sculptures as a fine example of "Japanese Modern" for the way Noguchi embraced tradition in his use of materials and techniques while at the same time creating sculptural forms

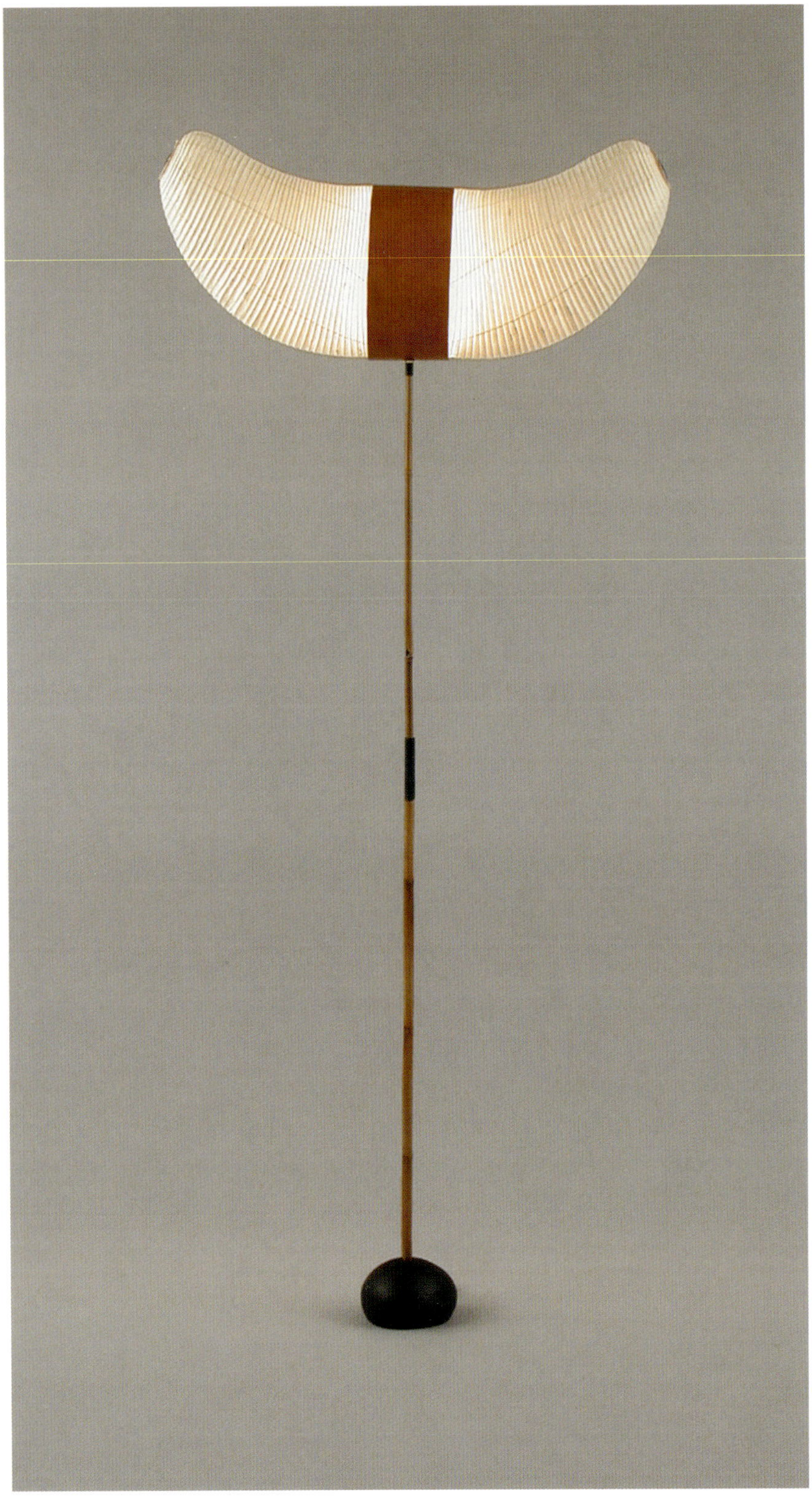

Isamu Noguchi, Akari BB3/33S, Washi paper, bamboo ribbing, and bamboo, ca. 1952 (stand, 1979), stand: 75 in., shade: 29 x 11 in., manufactured by Ozeki & Co., Ltd.

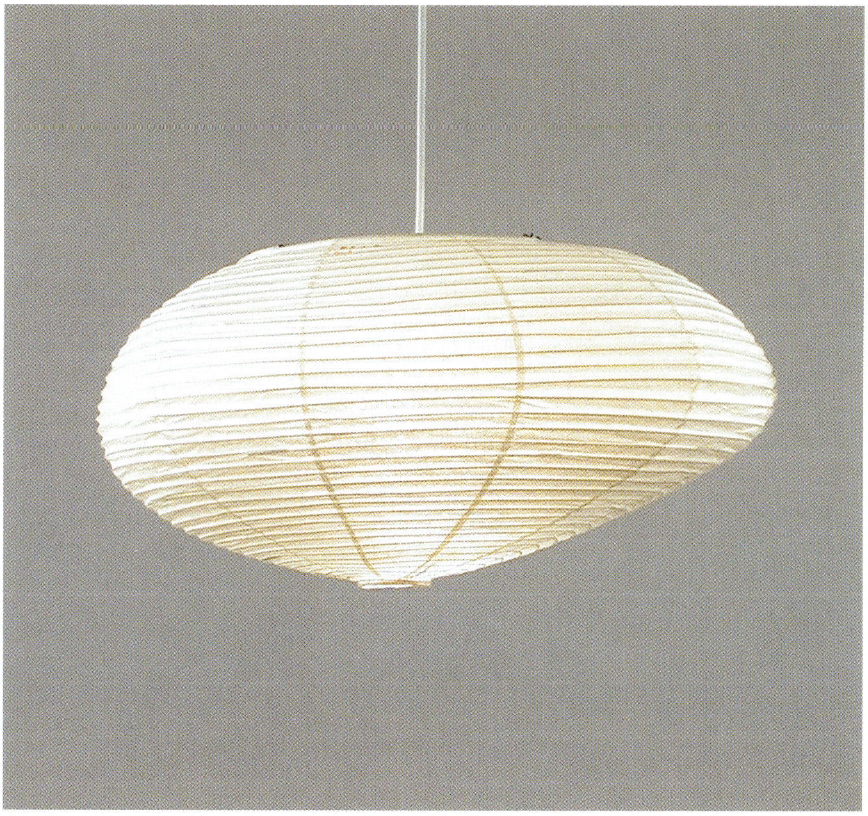

Isamu Noguchi, Akari 16A, Washi paper and bamboo ribbing, 1953, 20 x 19 x 9 1/2 in., manufactured by Ozeki & Co., Ltd.

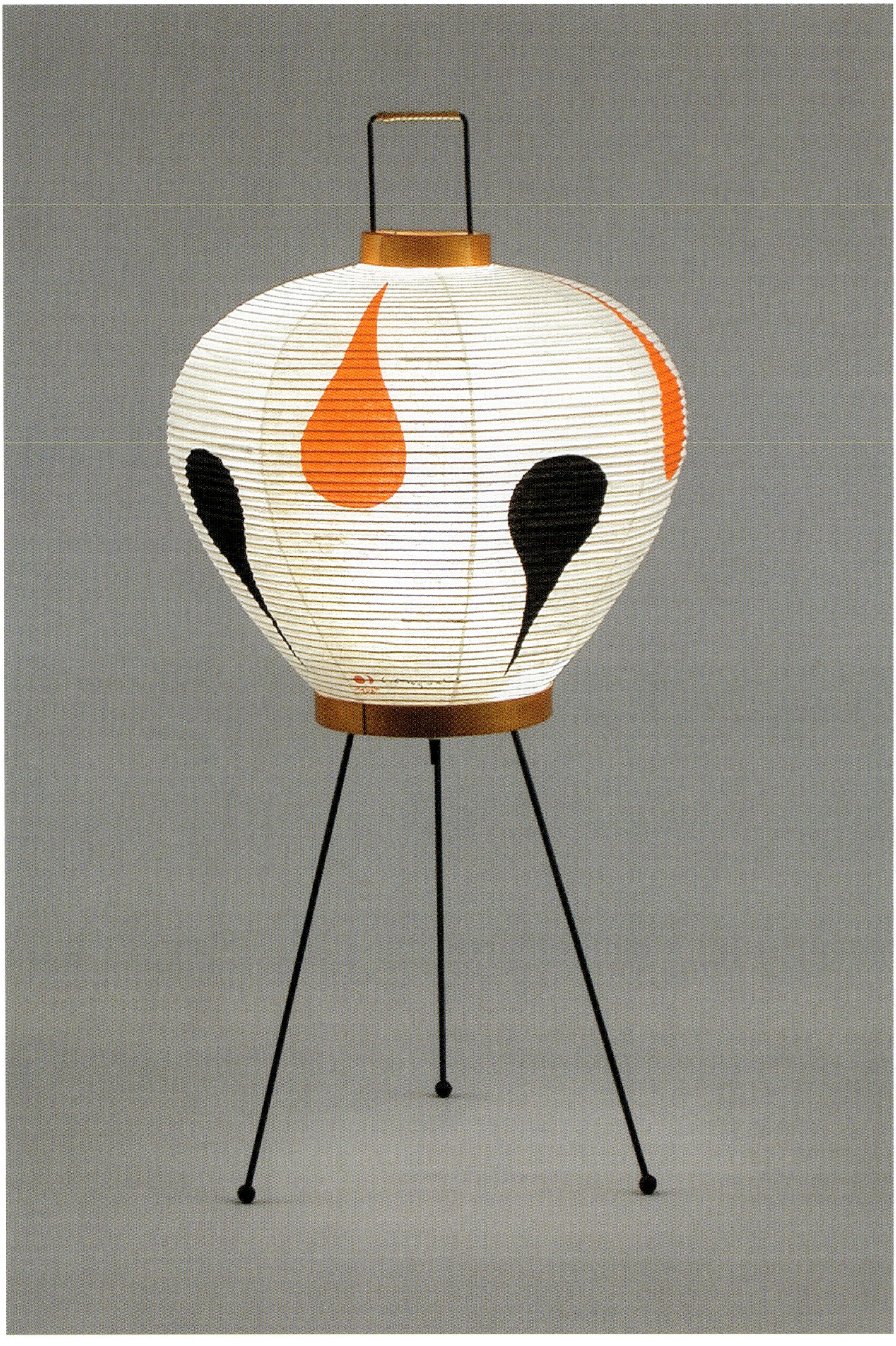

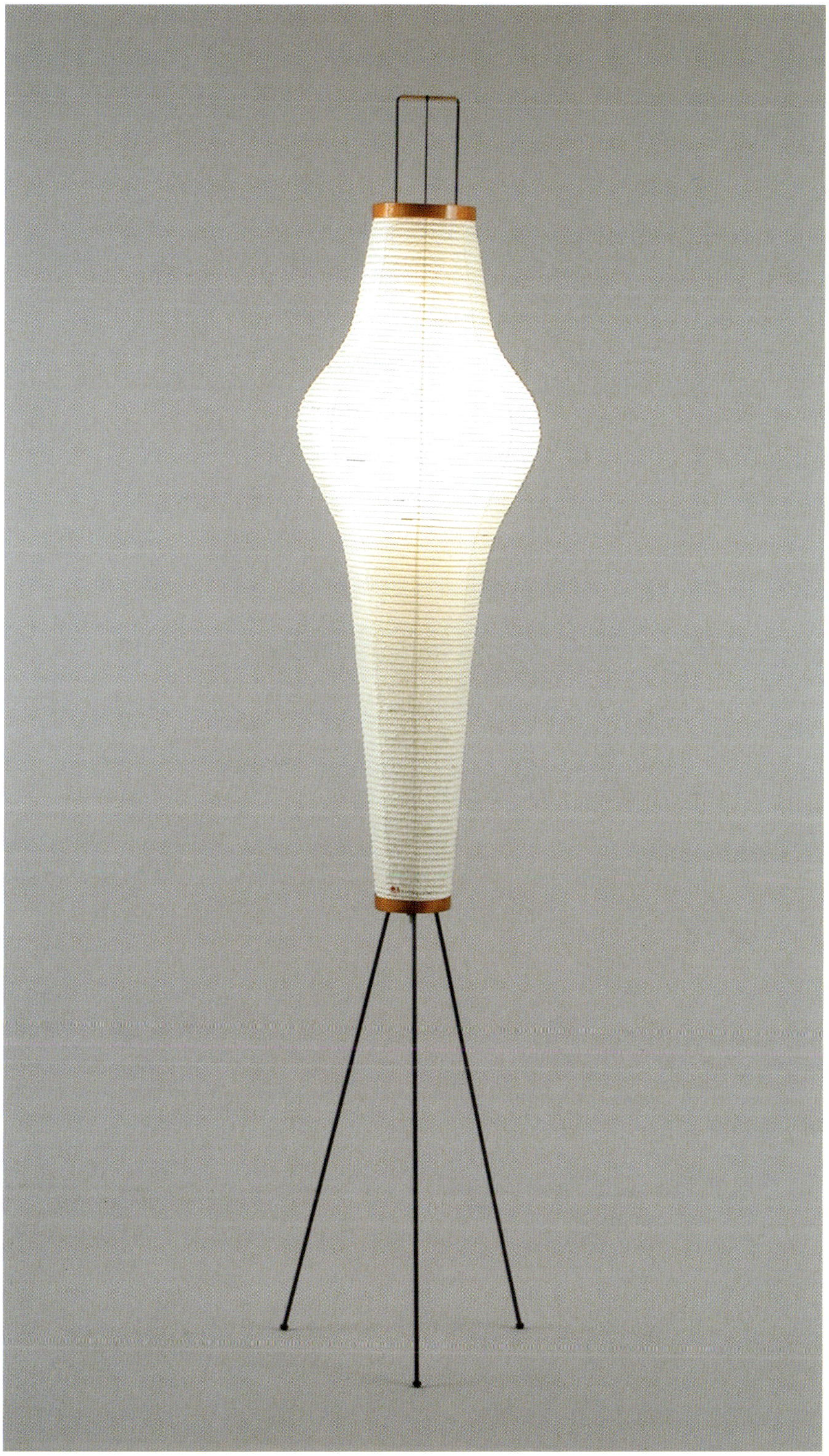

Above: Isamu Noguchi, Akari 14A, Washi paper, bamboo ribbing, and metal, 1952, height: 62 in., diameter: 20 in., manufactured by Ozeki & Co., Ltd. Opposite: Akari 3AD, Washi paper, bamboo ribbing, and metal, 1952, height: 22 in., diameter: 11 in., manufactured by Ozeki & Co., Ltd.

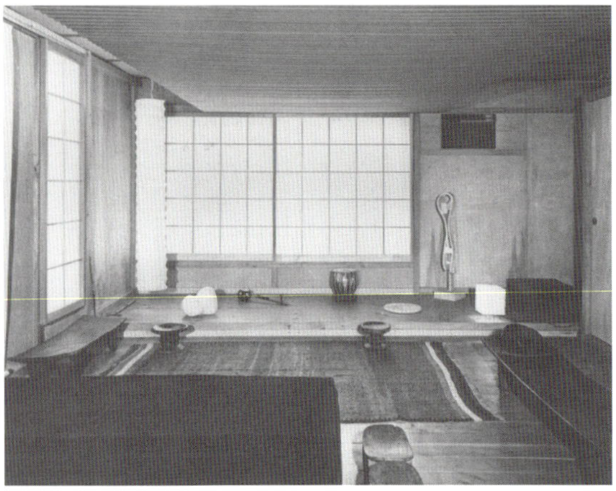

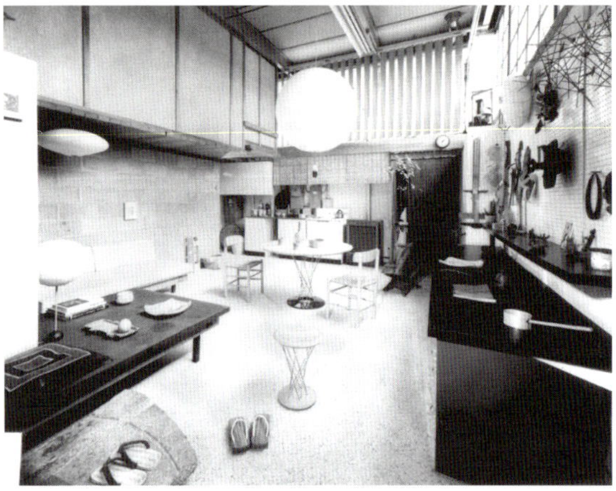

Top: Isamu Noguchi's Long Island City Studio, upstairs living area, ca. 1960s. Above: Long Island City Studio living area, ca. 1961.

to suit modern interiors.[7] Kenmochi suggested that Japanese Modern design products could be created by individuals of any nationality provided they were able to understand and regenerate Japanese cultural tradition with a modern sensibility. This involved a new intercultural attitude toward design while avoiding a false attempt to invent an exoticized Japanese style, which was evident among Japanese manufacturers in an effort to enhance foreign exports. Crucially, Kenmochi's concept of Japanese Modern design agreed with Noguchi's cross-cultural creative process; Noguchi traveled constantly throughout his life and adopted the materials at hand in each location, while nonetheless feeling attached to the nature of each material he worked with. "I always work with whatever medium is at hand," he wrote. "I don't believe in sticking to one medium. I'm afraid of its dominating me and becoming my trademark. . . . Of course, I am influenced by the material to the extent that when I work in heavy granite, I become heavy in thought and emotion; I don't think of flying."[8]

Noguchi took another direction in infusing Japanese design principles with foreign materials found outside of Japan. For instance, when he converted a building he bought in Long Island City, New York, into his residence and studio in the early 1960s, he used readily available, inexpensive materials such as plywood, pegboard, fiberglass, and Masonite. The overall outcome of his design was considered Japanese Modern. Writing in the *New York Times*, George O'Brien reported that the studio was "Japanese in feeling" despite the abundant use of Western materials that were "left natural, [so that] their plain and honest good looks contribute greatly to the understated style that is so typical of Japanese interiors."[9] In other words, understanding the essence of a traditional Japanese interior, Noguchi redefined it in the local context of New York. He proved that Japanese Modern design is not about the abundant use of authentic Japanese materials, but rather about the way in which materials are used.

With this cross-cultural design principle, Noguchi continuously redefined his design works throughout his career. A shift in his design approach can be traced to the Akari Light Sculptures, designed from 1951 until the end of his life. Akari have been shown in a variety of cultural, historical, and architectural contexts, and Noguchi periodically renewed their form, scale, and installation schemes. He began to design a folding lantern when he first visited Gifu, Japan, in 1951.[10] The project became Noguchi's life work, yet the idea of a light sculpture had begun to germinate as

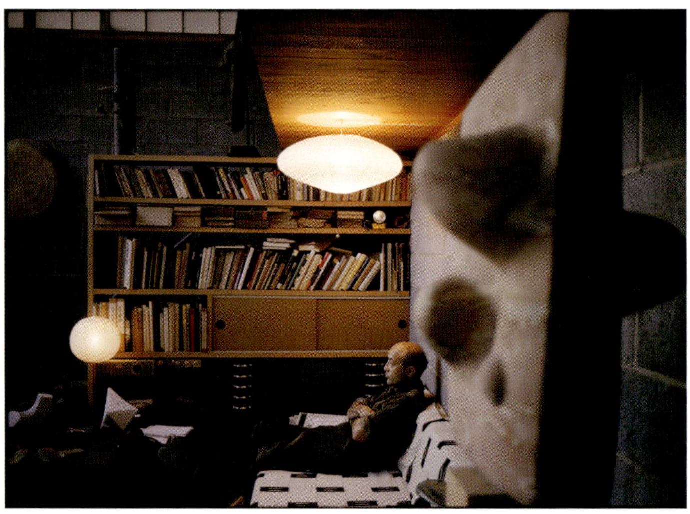

Storage Wall in Noguchi's Long Island City Studio, maple, 1962.

early as 1928 when Noguchi created *Power House*, a sculpture made of a neon tube and inspired by the Modernist, futuristic embrace of technological advancement. Almost two decades later, Noguchi made a series of "Lunars" linked to a Surrealist fantasy, an unfamiliar vision of suggested organic forms illuminated in space by their own mysterious light. In contrast, Noguchi designed Akari with a different mentality, paying homage to regional materials and to their convenient use in everyday life.

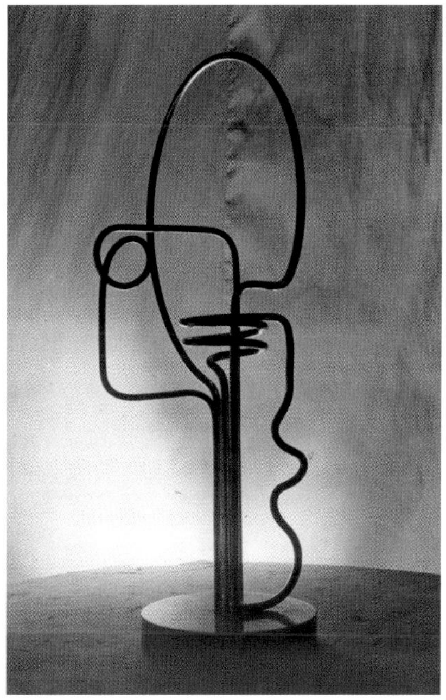

Isamu Noguchi, *Power House*, (lost), 1928.

His early designs of Akari were shown at the Museum of Modern Art, Kamakura, in 1952, and at the Chuo Koron Gallery in Tokyo in 1954. Various types of Akari were placed on tables and on the floor, and were installed on the walls and ceilings of the galleries. Traditional wooden rims are still visible in some of the Akari, but later Noguchi integrated the whole structure without a rim. The new design was achieved in part by using electric light in place of the traditional candle. In a 1954 article Noguchi described his passion for keeping tradition alive:

> I think that candles are more humanly attractive than electric lights, but there must be change, without which lanterns too must go, whatever their sentimental appeal. And to survive only as festive decorations is to deny lanterns a very real part of their intrinsic beauty, as well as their honorable tradition as a source of light.[11]

The relatively small size and varied placement of Akari made them practical for lighting domestic interiors, and reflected Noguchi's serious effort to innovate, creating everyday postwar Japanese products with a distinctive cultural identity.

While Akari attained some visibility in art gallery and museum spaces in Japan in the 1950s, as light sculpture Akari have been difficult to locate in the historical context of modern art in the United States. Initially they received attention as Good Design objects rather than as art objects. "Good Design" shows were organized by Edgar Kaufmann, Jr., at The Museum of Modern Art in New York between 1950 and 1955 in order to promote public awareness of sophisticated taste in everyday products.[12] The selection committees were chaired by Kaufmann and consisted of design-related professionals appointed each year such as Philip C. Johnson, Florence Knoll, Eero Saarinen, and Russell Wright. Noguchi's three-legged cylinder lamp with a translucent plastic shade was included in Kaufmann's 1950 Good Design campaign book titled *What is Modern Design?*.[13] Later Akari, listed as one metal and paper "floor lamp" costing $15 and two other "Table lamps"

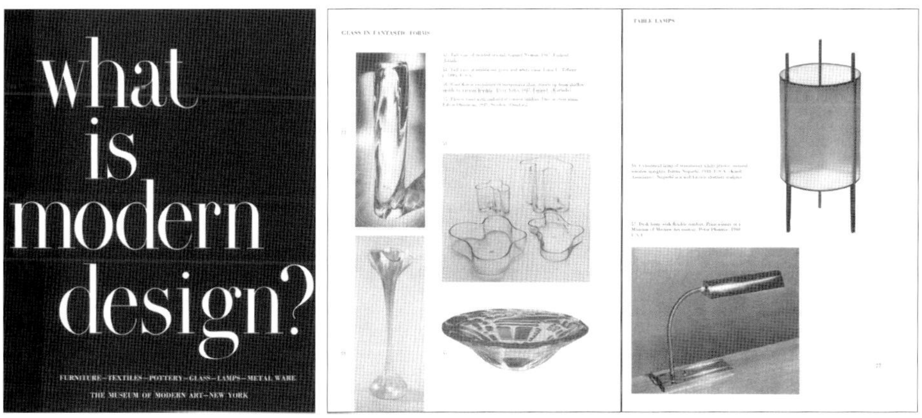

Edgar Kaufmann, Jr., *What is Modern Design?*, 1950, cover and pp. 26–27.

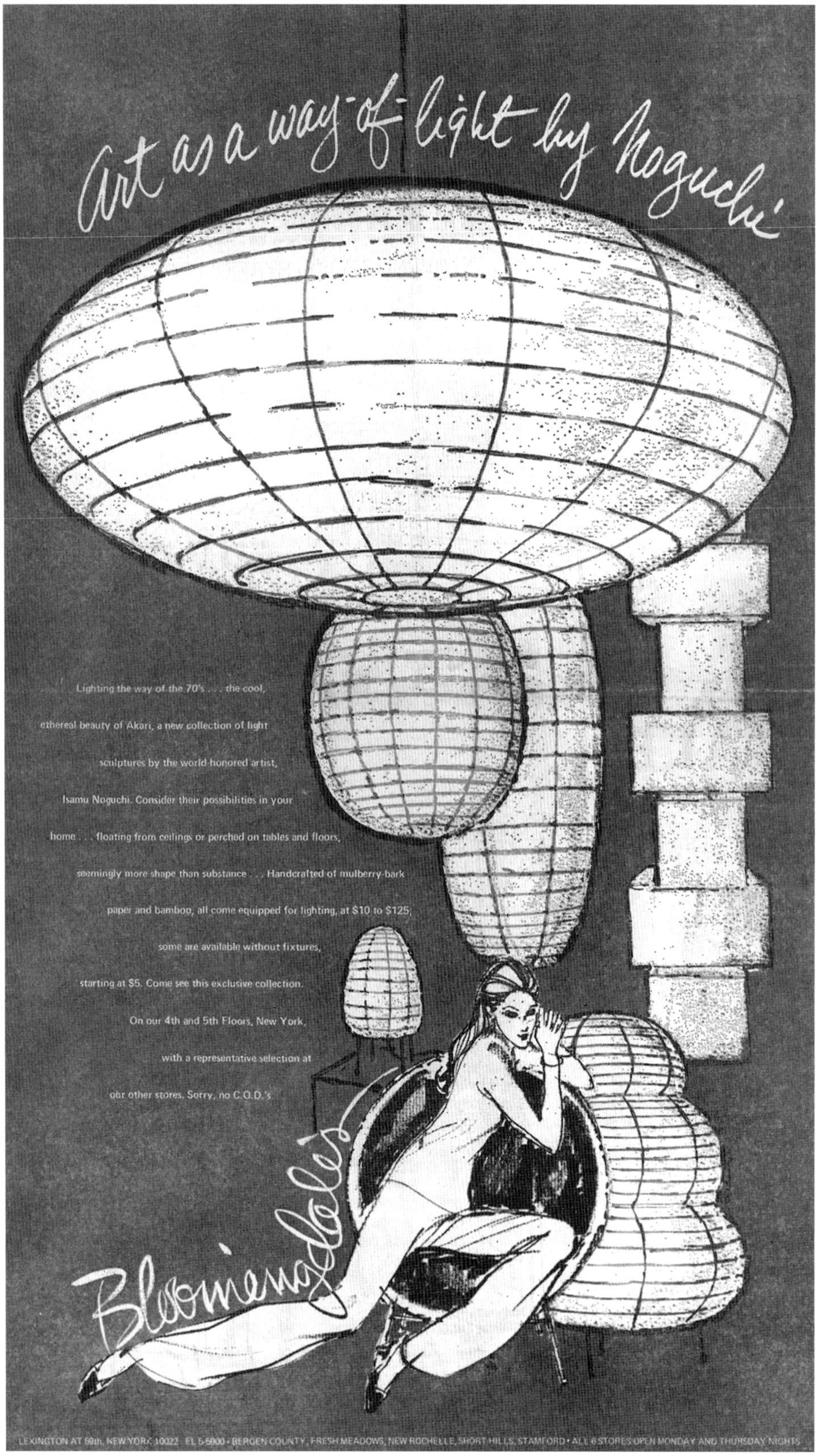

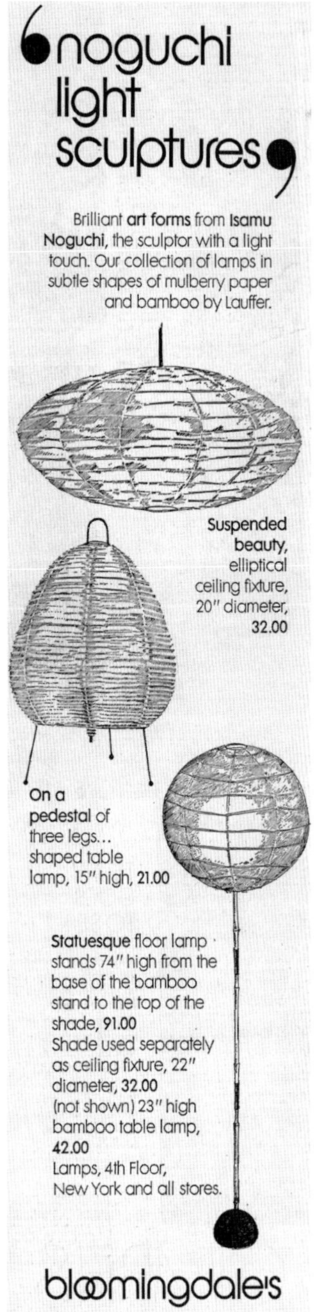

Above left and right: Bloomingdale's newspaper advertisements for Akari from the *New York Times*, 1970.

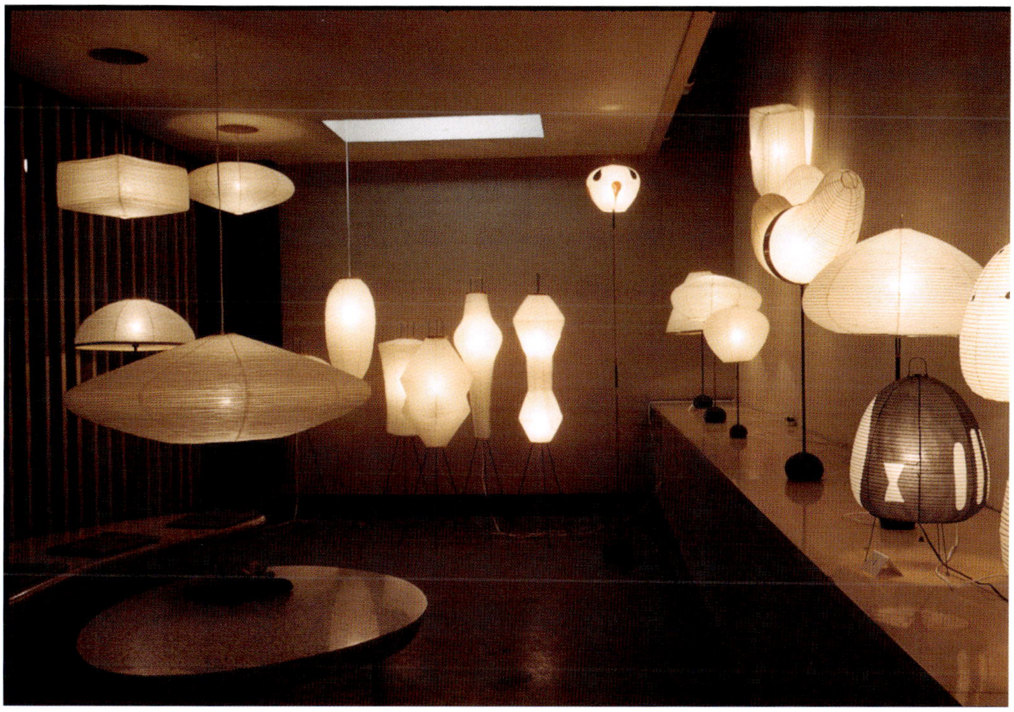

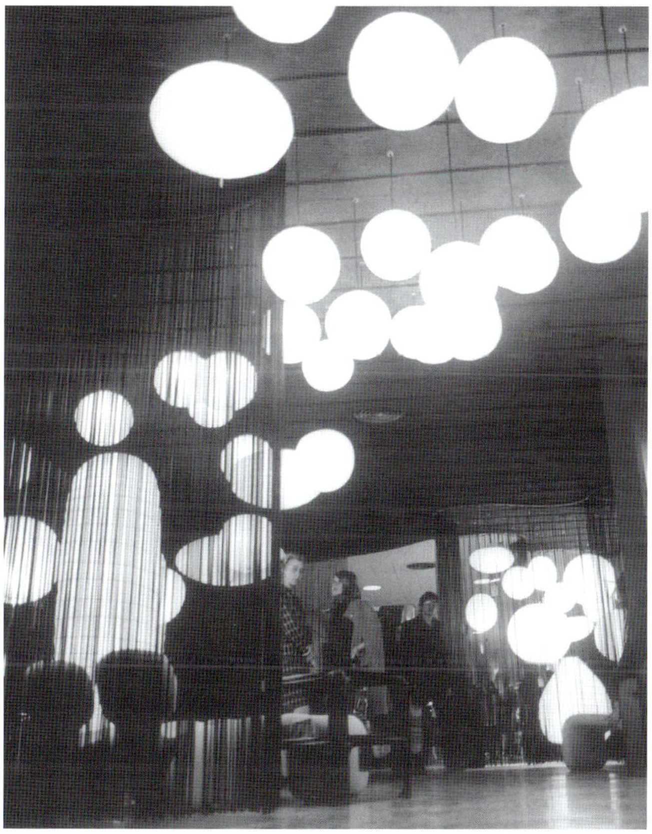

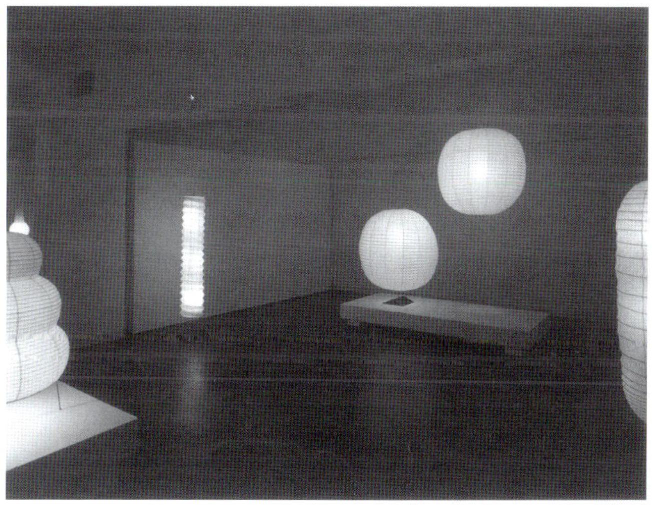

Clockwise from top: Installation view of Akari at the Chuo Koron Gallery, Tokyo, 1954; installation view of Akari at the Cordier and Ekstrom Gallery, New York, 1968; Bloomingdale's opening for Akari, 1970.

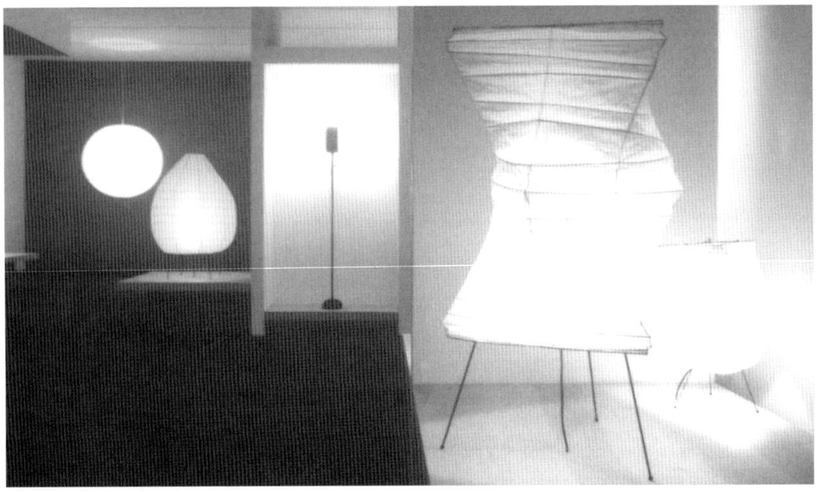

Installation view of the exhibition *Isamu Noguchi: Sculptor as Designer* at The Museum of Modern Art, New York, 1977.

costing $12.50 and $8.50, were chosen by the Good Design committee in 1954. They were made in Japan, packaged flat, and distributed by Bonniers in New York.[14] The department store Bloomingdale's agreed to distribute Akari, and publicized them as high-fashion interior objects through sleek advertisements in the *New York Times* in 1970.

A Noguchi retrospective was held at the Whitney Museum of American Art in 1968, and Noguchi increasingly established his artistic career through the late 1960s and 1970s. By that time, Akari had become icons of modern design. In an interview for *Art in America* in 1968, Noguchi pointed out that earlier in his career it was impossible for him to show a functional object in a gallery and call it art, but later it became acceptable for him to do so in the art world.[15] His emphasis on connecting art and life experience was developed before the war, and this same idea was received differently in the youthful counter-culture context of the 1960s. Noguchi was aware of the rising critique of the commodification of art among the younger generation of artists. "I see no difference in designing a piece of furniture or a piece of sculpture. Each is as valid. But, if you think making money taints art, I make more on my sculpture than on my furniture designs," Noguchi said to an interviewer in 1968.[16] This is a provocative and perhaps even cynical statement, but for Noguchi, Akari were not a mere commodity; they represented an artistic medium brought into everyday use.

Although Noguchi conceived of Akari as light sculpture, they have hardly been appreciated as such by Western audiences. This is largely due to cultural conventions, as Arthur Drexler, curator at The Museum of Modern Art, New York, explained in his talk on December 10, 1977, in conjunction with the show *Isamu Noguchi: Sculptor as Designer*. Drexler pointed out that fine art is more lofty than decorative art in the Western cultural tradition, whereas in Japan and the so-called "Orient" educated eyes respond to ceramics.[17] Movements such as the Bauhaus dealt with this discrepancy. However, despite the vanguard attempts of early modern art, Drexler noted that "we [Westerners] don't get the same feeling, perceptions" shared among cultures in the East.[18] In fact, the exhibition at The Museum of Modern Art was curated by J. Stewart Johnson, curator of design from the Department of Architecture and Design, and its primary focus was Akari with the addition of furniture and industrial products, such as Noguchi's Glass-Top Coffee Table and his Radio Nurse, an

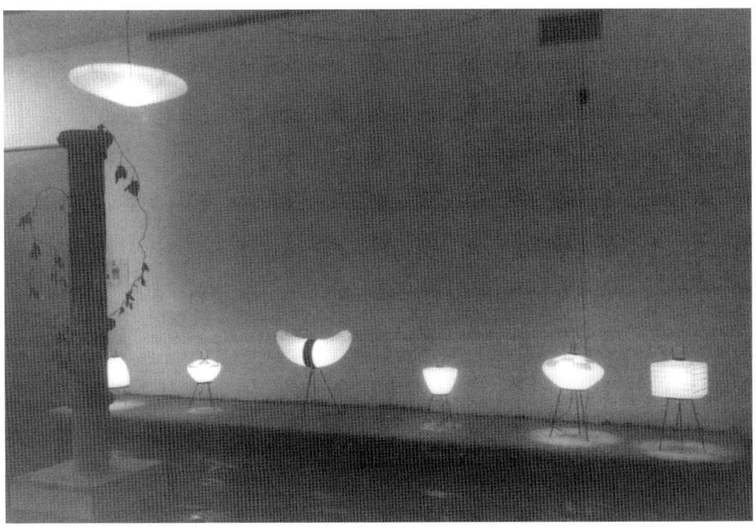

Installation view of Akari at the Museum of Modern Art, Kamakura, 1952.

early design of an audio baby monitor. Akari were treated as lamps in line with the Good Design legacy rather than as sculpture in their own right.

Meanwhile, Akari Light Sculptures, which originated in Japan, were indeed taking off from their own roots in the 1950s, and Noguchi continued to develop them as an art form in the following decades. He designed Akari on a more ambitious scale while emphasizing the aesthetic quality of their light and their bold presence in space. "They're like balloons except they don't collapse," Noguchi said of Akari in 1970, "and the bigger they are, the better. A volume of light brings the sun into your life. . . . They're not material clumps but emanations of light vibrations."[19] He could successfully show Akari at art galleries once the time was right to do so in the late 1960s. Compared to the practical Akari designs Noguchi presented at the Chuo Koron Gallery in 1954, Akari were treated more as sculptures at the Cordier and Ekstrom Gallery in New York in 1968. The large Akari displayed at the gallery were installed in a spacious setting on low pedestals. By showing Akari as sculpture at galleries, Noguchi challenged the hierarchical system of Western cultural conventions.

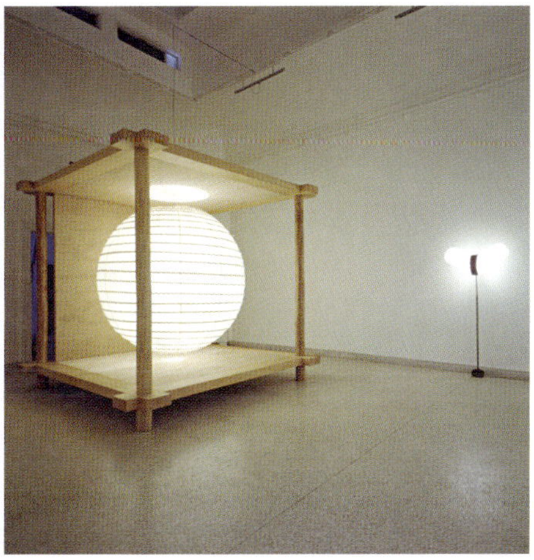

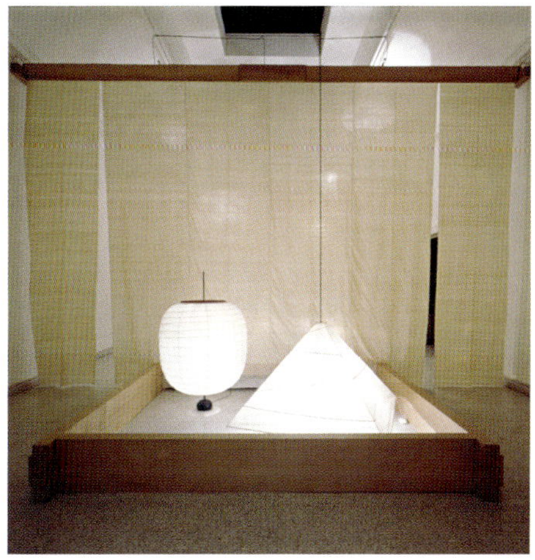

Installation views of Akari in the United States Pavilion at the 1986 Venice Biennale.

Perhaps Noguchi's most ambitious presentation of Akari took place at the Venice Biennale in 1986, when he was chosen to represent the United States at this international art competition. Visitors to the United States Pavilion were confronted by a huge Akari globe framed by wooden columns, along with other Akari displayed on the floor and surrounded by a low square structure. The geometric elements set around the Akari served a function similar to that of pedestals in the Western sculptural tradition, separating them from their use and giving their gentle light more volume and strength. The monumental Akari globe in particular generated an aesthetic encounter different from the intimate presence of its light in everyday life. Akari at the Venice competition lost some of their nature as objects of ephemeral beauty, lightness, and fragility when Noguchi redefined them as pieces of art on a grand scale with a physical presence comparable to Western sculpture. Nevertheless, Akari remain functional objects with a Japanese sensitivity to the warmth of subtle light, and this created an odd discrepancy in the context of the cold, white, dignified neoclassical architecture of the United States Pavilion.

In a way, Noguchi's ambivalent cultural identity was revealed in his installation of Akari at the Venice Biennale. He was received as half-Japanese, not only because of his background, but also because of his lifelong devotion to Japanese aesthetics as both an artist and a designer. It was unfortunate that some criticism of Akari at the Biennale did not go beyond a preconceived notion

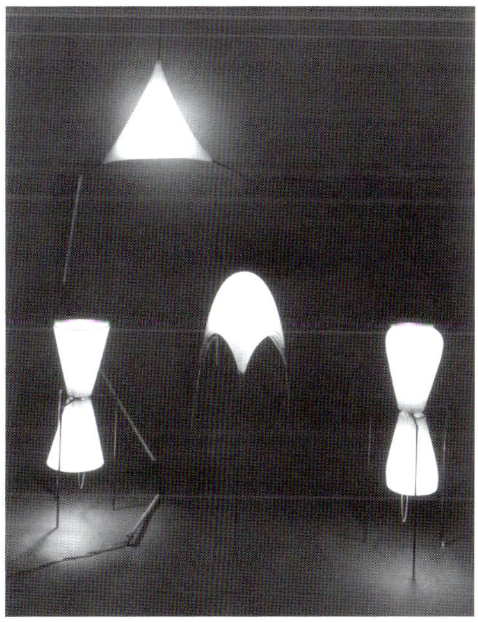

Isamu Noguchi, Lunars, 1944, from *LIFE* magazine, March 10, 1952.

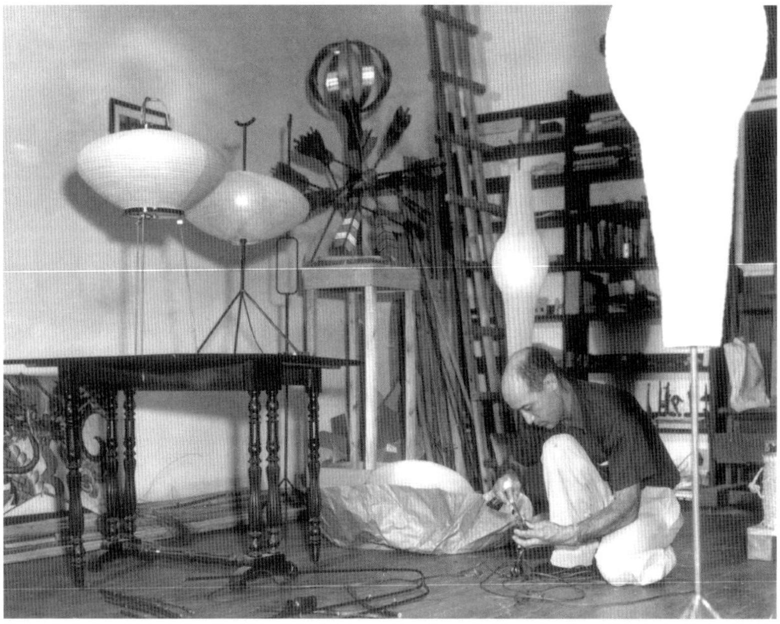

Isamu Noguchi at work on Akari Light Sculptures, ca. mid- to late 1950s.

about a so-called "oriental" artist stressing his strong ties to cultural tradition and ignored the present context of his works. Noguchi posed a question about the increasingly irrelevant idea of an artistic competition among nations by manifesting his own cross-cultural identities through Akari.

Regardless of cultural or national boundaries, Akari encourage the viewer's contemplation of light. They provoke an absorption into emanating light, just as Noguchi often felt absorbed in the materials he worked with locally. In this private realm of looking at light, Akari do not have to be defined as art objects or as functional objects. Ultimately, they reside anonymously in the mind at the moment when the viewer is immersed in their light. One aspect of Isamu Noguchi's contribution to modern design is this invitation to accept things as anonymous, breaking down the barrier between self and other. His cross-cultural design principle for coping with cultural differences is about working locally with a global perspective. There will continue to be friction between diverse cultures in contemporary society, but Akari suggest a subtle solution to this friction. While illuminating the imaginative space of each individual, they alert us to the possibility of coexistence with other cultures, even as they remind us to keep a respectful distance from others immersed in this common light.

NOTES

My argument is focused on the cultural aspects of Noguchi's design. Hence, I do not investigate Noguchi's design works associated with Surrealism, Constantin Brancusi (1876–1957), or Buckminster Fuller (1895–1983). The subject of Noguchi as a Surrealist-affiliated artist and designer was explored in an exhibition catalogue, Brooke Kamin Rapaport et al., *Vital Forms: American Art and Design in the Atomic Age, 1940–1960* (New York: Brooklyn Museum of Art in association with Harry N. Abrams, 2001), and Brancusi's influence on Noguchi was discussed in Anna Chave's essay, "Brancusi and Noguchi: Towards a Larger Definition of Sculpture," in *Isamu Noguchi: Sculptural Design* (Weil am Rhein: Vitra Design Museum, 2001), pp. 26–65. This latter text was an inspiration for my own essay. A close friendship and mentorship between Noguchi and Fuller was explored by Shoji Sadao in *Best of Friends: Buckminster Fuller and Isamu Noguchi* (NY: The Isamu Noguchi Foundation and Garden Museum, 2006).

1. Isamu Noguchi "Modern Trends in Interior Furnishings" *Industrial Art News,* vol. 18, no.10 (October 1950): 334, translated from Japanese to English by the author.
2. Isamu Noguchi, Tokyo, to Bernard Bergstein, New York, May 16, 1950. Copyright 2007 The Isamu Noguchi Foundation and Garden Museum, New York.
3. Isamu Kenmochi, "Isamu Noguchi As I See Him," *Industrial Art News,* vol. 18, no. 10 (October 1950): 21.
4. Isamu Noguchi, *Art and Architecture* (November 1950): n.p.
5. Shuzo Takiguchi, "A Strange Art Journey (Isamu Noguchi)," in *Noguchi* (Tokyo: Bijutsu Shuppan-sha, 1953): n.p.
6. Isamu Noguchi, Tokyo, to Bernard Bergstein, New York, postal stamp April 24, 1951. Copyright 2007 The Isamu Noguchi Foundation and Garden Museum, New York.
7. Isamu Kenmochi, "Japanese Modern & Japonica Style: Two Roads to Put Japanese Industrial Arts on Foreign Export," *Industrial Art News*, vol. 22, no. 9 (September 1954): 7.
8. Katherine Kuh, "An interview with Isamu Noguchi," *Horizon Magazine*, vol. 11, no. 4 (March 1960): n.p.
9. George O'Brien, "Factory into Home," *New York Times Magazine*, April 8, 1962.
10. See an essay by Takuya Kida on the development of Noguchi's design of Akari and their critical reception in Japan in *AKARI: Light Sculpture by Isamu Noguchi* (Tokyo: The National Museum of Modern Art, 2003): 8–23.
11. Isamu Noguchi, "Japanese Akari Lamps: Light, Translucent Sculpture by a Great Sculptor," *Craft Horizons*, vol. 14, no. 5 (1954): 18.
12. See a critical insight about Good Design shows at The Museum of Modern Art, New York, in the essay "The 'Good Design' Syndrome" by Arthur Pulos in *The American Design Adventure, 1940–1975* (Cambridge, MA: MIT Press, 1988): 110–121.
13. Edgar Kaufmann, Jr. *What is Modern Design?* (New York: The Museum of Modern Art, 1950). The book was translated from English to Japanese by Tsutomu Ikuta and published by Bijutsu Shuppan-sha in 1953.
14. The Museum of Modern Art, New York, and Merchandise Mart, Chicago, *Good Design* (New York: The Museum of Modern Art, 1950–1955): n.p.
15. John Gruen, "The Artist Speaks: Isamu Noguchi," *Art in America*, vol. 56, no. 2 (March–April, 1968): 30.
16. Carolyn Hall, "Noguchi Discusses Sculpture, Design," *Detroit* (January 1968): n.p.
17. Arthur Drexler, introduction to the artist's talk in conjunction with the exhibition *Isamu Noguchi: Sculptor as Designer*, presented at Titus Auditorium at The Museum of Modern Art, New York, December 10, 1977, Sound Recording of Museum-Related Events, 77–81, The Museum of Modern Art Archives, New York.
18. Ibid.
19. Isamu Noguchi, quoted in Susan Bogers, "Noguchi: Beauty Is Not Forever," *New York Post*, Tuesday, April 21, 1970: 6.

I thank Amy Hau, Administrative Director, and Heidi Coleman, Photo Archivist, at The Noguchi Museum in New York for extensive access to the Museum's archives.

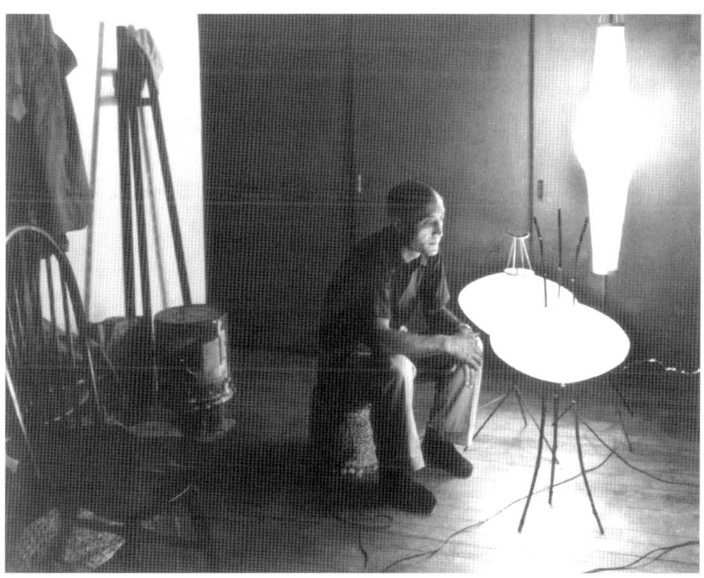

Isamu Noguchi with Akari Light Sculptures, ca. 1950s.

A CHRONOLOGY OF AKARI

NINA MURAYAMA

Originally compiled by Takuya Kida in *AKARI: Light Sculpture by Isamu Noguchi* (Tokyo: The National Museum of Modern Art, 2003). Revised with additional citations by Nina Murayama.

1950 **May, Noguchi visits Japan for the first time after the end of World War II.**

August 1, Noguchi spends two weeks making sculpture and furniture at the Industrial Arts Research Institute, Tokyo.

August 18–27, Mainichi Newspapers Co., Ltd., sponsors an exhibition, *The Works of Isamu Noguchi*, at Mitsukoshi Department Store in Tokyo.

1951 **March, Noguchi returns to Japan to design a garden for the Reader's Digest Building by architect Antonin Raymond.**

June, Noguchi stops in Gifu to see Ukai on his way to Hiroshima to design two railings, known as "Tsukuru" and "Yuku," for the Hiroshima Peace Park. Noguchi visits the lantern factory Ozeki and Co. Ltd., and begins to design Akari.

August, completion of *Shin Banraisha* at Keio University.

1952 **September 23–October 19, the exhibition *Isamu Noguchi* is on view at the Kamakura Museum of Modern Art, and Akari are shown for the first time.**

October, Akari are exhibited and sold well at the Takashimaya Department Stores in Tokyo and Osaka.

Model numbers of Akari designed and produced: 1A through 14A.

1953 **Bonniers, a store of modern design in New York, begins to sell Akari.**

Noguchi applies for a patent in Japan for "Stretcher and Support for Japanese Lanterns." In the following decades, Noguchi will continue to patent Akari designs in the United States and Japan as he develops them throughout his career.

Model numbers of Akari designed and produced: 15A through 18A.

1954 **Noguchi applies for a patent in the U.S. for "Stretcher and Support for Japanese Lanterns."**

August—September, exhibition of Akari at Chuo Koron Gallery, Tokyo.

Wohnbedarf AG in Zürich begins to sell Akari in Switzerland and West Germany.

Akari are selected for exhibition by the Good Design Committee, chaired by Edgar Kaufmann, Jr., at The Museum of Modern Art, New York.

Model numbers of Akari designed and produced: 21A, 24A, 25A, 45A, 55A, A, B, E.

1955	April 20, opening of Akari exhibition at Bonniers, New York. Model number of Akari designed and produced: 26A.
1956	Wohnbedarf AG begins to distribute Akari in Austria, Denmark, and Sweden. By this time, patents are granted in the United States and Japan, and imitations of Akari are internationally ubiquitous. Model number of Akari designed and produced: 30A.
1957	Model number of Akari designed and produced: D.
1958	Akari are sold at the Good Design shop in Matsuya Department Store, Tokyo. Steph Simon in France receives samples of Akari after a commercial treaty was signed between Japan and France in 1957, allowing the import and export of commodities; Simon becomes the exclusive representative of Akari in France.
1959	Wohnbedarf AG and Noguchi sign an agreement to distribute Akari in Austria, Benelux, Germany, Italy, and Switzerland.
1960	Model number of Akari designed and produced: 23A.
1961	Akari are shown at the Frankfurt Fair. First production of Akari metal stand. Model numbers of Akari designed and produced: 22A, 75A, 2X, 10T, H, I, J, K, L, T2 (Stand), T3 (Stand).
1962	April, exhibition of Akari at Bonniers in New York. First production of Akari bamboo stand. Model numbers of Akari designed and produced: 41S (Stand), 42M (Stand), 43L (Stand).
1963	With approval from Wohnbedarf AG, Hans W. Seleger in Switzerland expands the Akari market in Europe. Noguchi is ambivalent about winning the popular market, and is committed to maintaining the high quality of Akari. First production of the asymmetrical Akari 55D (D series) using bamboo reed. Model numbers of Akari designed and produced: 55D, 3X, 9T, T1 (Stand).
1964	April, exhibition of Akari at Bonniers, New York. August 14–September 2, exhibition of Akari in the Design Gallery at Matsuya Department Store, Tokyo, organized by the Design Committee. Model numbers of Akari designed and produced: 30D, 45D, 120A.
1966	Noguchi establishes the Akari Foundation in New York to foster artistic exchange between Japan and the United States.

1967	**Model number of Akari designed and produced: 37D.**
1968	**December 4–January 4, 1969,** *Shapes of Light*, **an exhibition including newly designed Akari, opens at Cordier and Ekstrom Gallery, New York.**
1969	**Yamagiwa Corporation begins exclusive distribution of fifteen types of "New Akari" (N series).** **October 15–25, the exhibition** *New Akari* **opens on the sixth floor of Yamagiwa's Tokyo branch.** **Akari are displayed at the Steph Simon Gallery in Paris.** **Akari Associates (Isamu Noguchi, Priscilla Morgan, and Shoji Sadao) is formed in New York to manage Akari distribution in the U.S. and abroad.** **First production of P series Akari without bamboo ribbing.** **Model numbers of Akari designed and produced: 1N, 4N, 20N through 24N, 30N through 34N, 10P, 10N, 12N through 14N, 30P, 31P.**
1970	**April 15, Bloomingdale's introduces Akari, installed in the center aisle on the 5th floor of its New York store.** **An Akari with a two-meter diameter is exhibited at Expo '70 in Osaka, Japan.** **Hans W. Seleger signs an agreement to sell Akari in selected European countries; the agreement with Bonniers in New York becomes nonexclusive.**
1971	**First production of Akari with distorted shapes (F series).** **Model numbers of Akari designed and produced: 60D, 75D, 50EN, 70EN, 95EN, 30F, 40F, 50F, 55F, 60F, 70F, 125F, 40DL, 60DL, 5N, 45X, 40XP, 31NW, 33NW, 25N, 26N, 27N, 45XL, 70XL, 90XL, 40S (Stand), 20S (Stand), 19S (Stand).**
1972	**Model number of Akari designed and produced: 80DL.**
1973	**January–February 4, Akari are included in the exhibition** *Art is Meaning–Art in Use* **at Wykeham Rise School, Washington Art Gallery, CT.** **Model numbers of Akari designed and produced: 1P, 2P, 33X.**
1974	**H. E. Lauffer Co. in New Jersey becomes a distributor of Akari in the U.S. (until 1977).** **Model numbers of Akari designed and produced: 100D, 95F.**
1975	**Model number of Akari designed and produced: 1X.**
1976	**Model numbers of Akari designed and produced: 5X, L1 through L8.**

1977	**November 11–January 8, the exhibition** *Isamu Noguchi: A Sculptor as Designer* **is on view at The Museum of Modern Art, New York, and many Akari are shown.** **George Kovacs Lighting, Inc., New York, agrees to distribute Akari in the U.S. and Canada (continuing to ca. 1984).** **Model numbers of Akari designed and produced: X1, V1, V2.**
1978	**September 30–October 27, an exhibition of Akari at the University of Hawaii, Honolulu, is on view, and travels to Chicago and Los Angeles.** **Model numbers of Akari designed and produced: X3, 30DL, 38DL, 75DL, 30XN, 45XN, 70XN, YP2, 30DD, 55DD, 75DD, 30FF, 55FF, 70FF.**
1979	**August 4–September 30, Akari Model #K4 is shown in** *Recent Acquisitions: Architecture and Design* **at The Museum of Modern Art, New York.** **First production of BB stand made of bamboo.** **Model numbers of Akari designed and produced: S1, S2, YP1, J1, BB1 (Stand), BB2 (Stand), BB3 (Stand).**
1980	**November–December, the exhibition** *Isamu Noguchi and Teshigahara Hiroshi* **is on view at Sogetsu Kaikan in Tokyo. Many Akari are shown.** **Model numbers of Akari designed and produced: YA1, YA2, 2N, K1.**
1981	**Model numbers of Akari designed and produced: XP1, JP.**
1982	**October 12–30, the exhibition** *Isamu Noguchi: Light Sculpture* **is on view at Kasahara Gallery, Osaka.** **October, Noguchi receives the International Culture Design Grand Prix from Japan Inter-Design Forum.** **Model number of Akari designed and produced: 40FF.**
1983	**April, the Isamu Noguchi Garden Museum in New York opens by appointment; Akari are featured in one of the museum galleries.**
1984	**September, Akari (UF series) are exhibited and sold at** *Japan* **at Bloomingdale's, New York.** **November 17–30, the exhibition** *Isamu Noguchi: Akari, Stone, and Terracotta* **is on view at Sogetsu Kaikan in Tokyo. Many Akari (UF series) are shown.** **Akari-Gemini, Santa Barbara, CA, begins to distribute Akari and to market them as art in the form of light sculpture.** **Model numbers of Akari designed and produced: UF1-C, UF1-O, UF1-FF, UF1-XN, UF2-H, UF2-U, UF2-FF, UF3-H, UF3-U, UF3-Q, UF3-L6, UF4-L5, UF5-32N, UF5-33NW.**

1985 **February 9–20, the exhibition** *Isamu Noguchi: Space of Akari and Stone*, **organized by the Seibu Museum of Art, is on view at the Yurakucho Art Forum in Tokyo, with installation design by Arata Isozaki.**

May 15–June 14, selections of Akari are shown in the exhibition *Light of Form, Paper and Steel* **at the Flushing Council on Culture and the Arts, Flushing, New York.**

September 19–February 16, 1986, Akari are shown along with Noguchi's Coffee Table (1944) and Chess Table (1947) in *High Style: Twentieth Century American Design* **at the Whitney Museum of American Art, New York.**

Model numbers of Akari designed and produced: UF1-H, UF1-S, UF1-Y, UF2-Y, UF2-X, UF2-31N, UF2-33N, UF3-S, UF3-DL, UF3-XN, UF4-L8, UF4-L9, UF4-J1, UF4-31N, UF4-33N, UF5-L4, UF5-31NW.

1986 **June 29–September 28, Noguchi represents the United States at the 42nd Venice Biennale. Newly designed and unique Akari are exhibited at the United States Pavilion.**

Model numbers of Akari designed and produced: UF4-L6, UF4-L10.

1988 **December 30, Isamu Noguchi dies at New York University Hospital at age 84, following a brief illness.**

This chronology of Akari model numbers is based on information provided by Mr. Hidetaro Ozeki; the exhibition catalogues *Isamu Noguchi* **(Tokyo National Museum of Modern Art, 1992) and** *Isamu Noguchi and Rosanjin Kitaoji* **(Sezon Museum of Modern Art, 1996); and sources in the Isamu Noguchi Archives in New York.**

From 1951, Noguchi worked exclusively with Ozeki & Co., Ltd., a lantern factory in Gifu, Japan, which continues to manufacture Akari under an agreement with The Noguchi Museum, including the rights to foreign distribution in consultation with The Noguchi Museum.

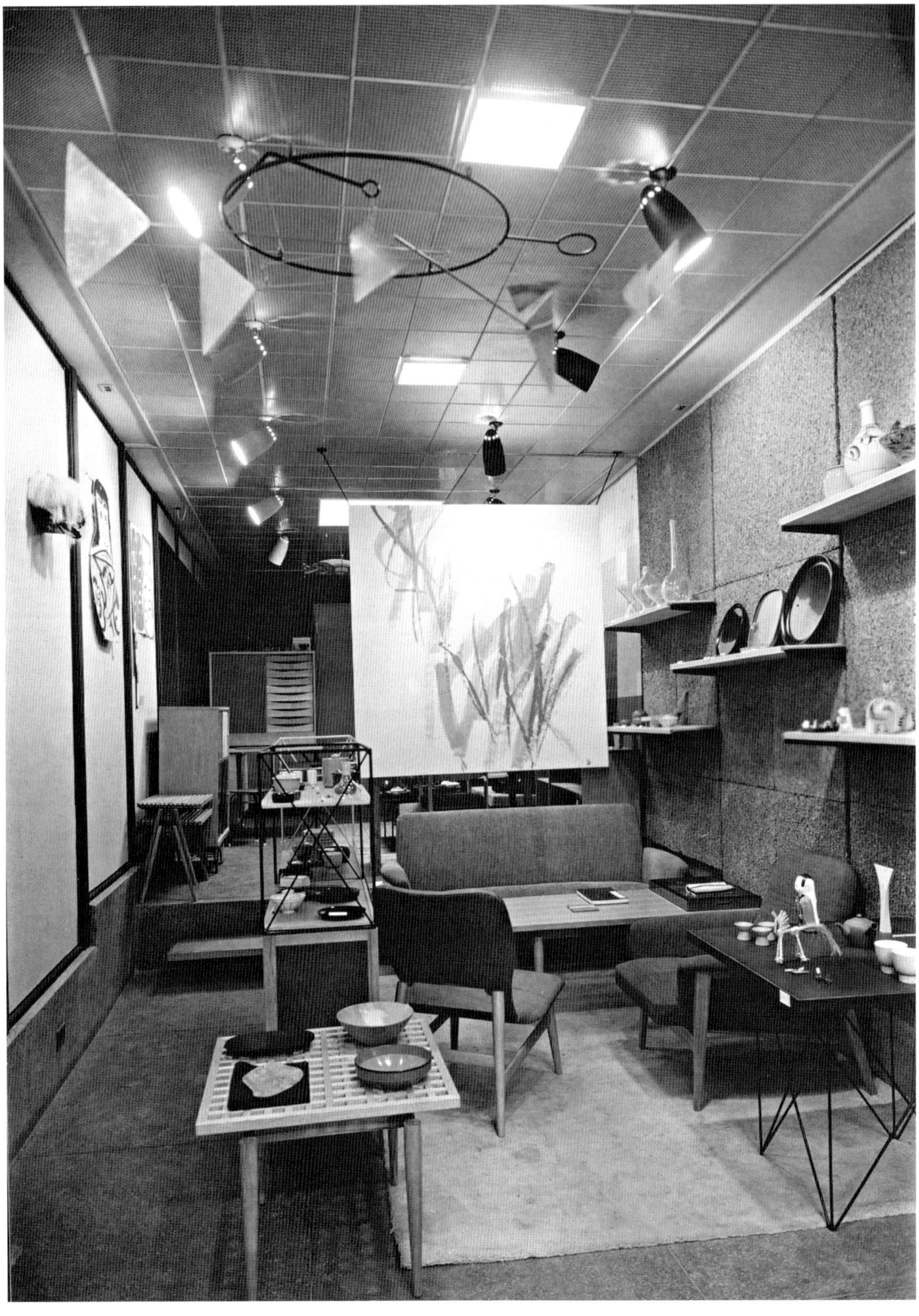

Unknown photographer, Living Art Showroom, ca. 1956–1958. Kenmochi established his own design firm, Kenmochi Design Associates, in the Aoyama district of Tokyo in 1955. High-quality design products were displayed and sold from the company's first-floor Living Art Showroom from 1956 to 1958. The concept was inspired by the "Good Design" series of exhibitions organized by The Museum of Modern Art, New York, from 1950 to 1955.

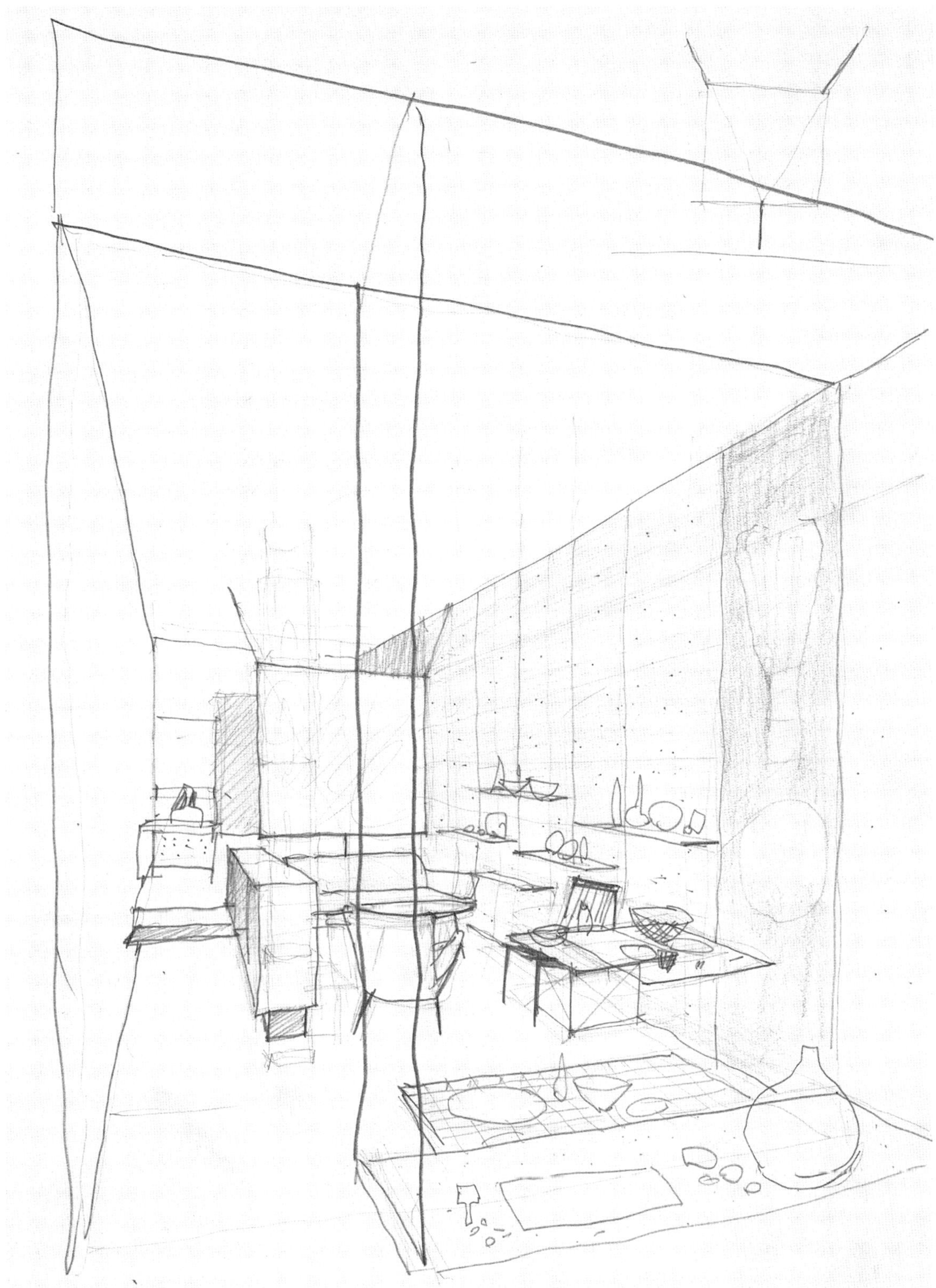

Isamu Kenmochi, Living Art Showroom, pencil on paper, 1955, 11 5/8 x 8 1/4 in.

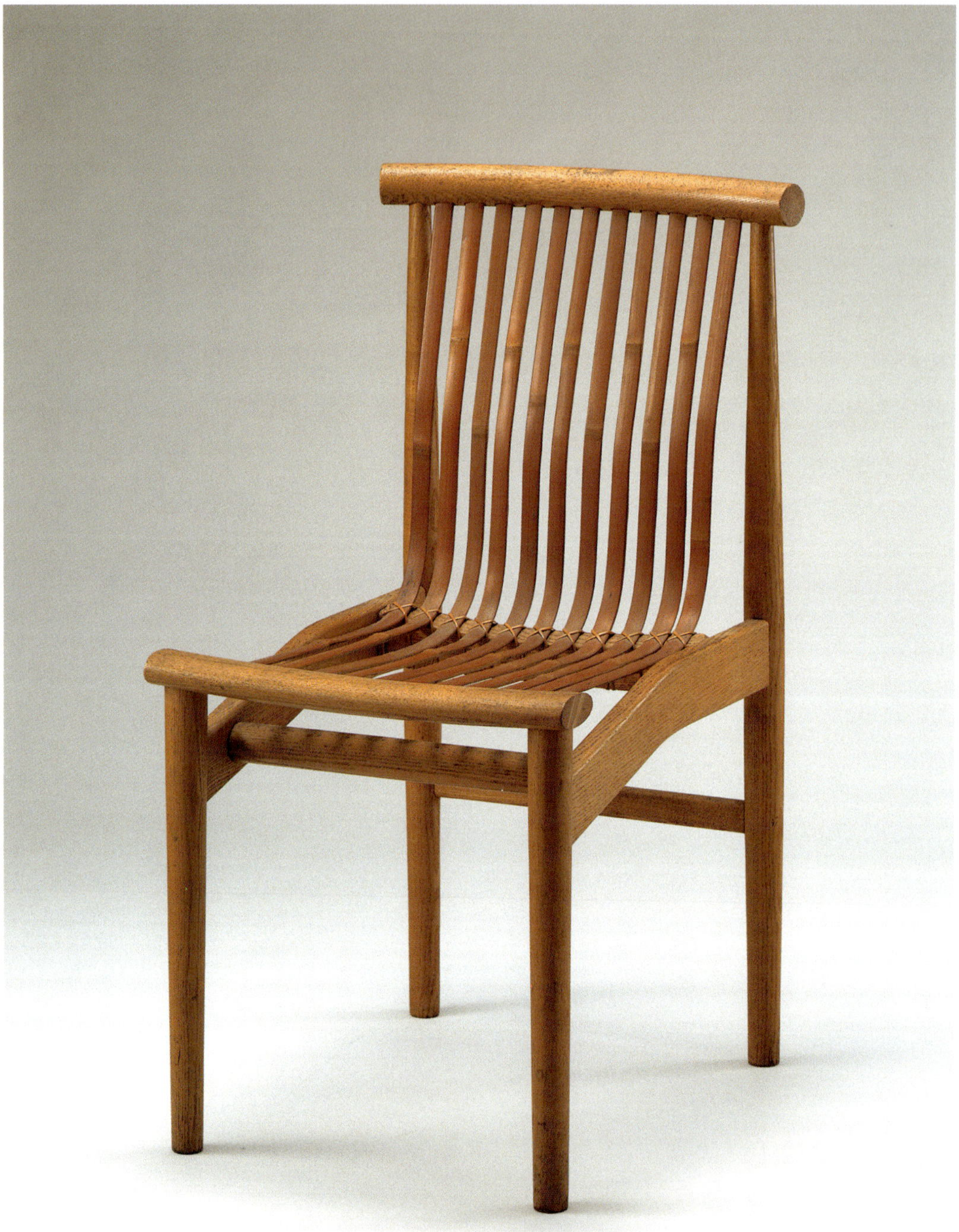

Isamu Kenmochi, Bamboo Chair, Shioji, bamboo, 1954, 32 3/16 x 16 9/16 x 20 1/4 in. Following his collaboration with Noguchi on the Bamboo Basket Chair, Kenmochi continued to explore and develop the fabrication of bamboo furniture, aware of the increasing demand for bamboo household items in the postwar U.S. market. Seeking to reinvent traditional Japanese bamboo techniques, he experimented with various chemical processes in order to achieve his concept of uniting time-honored Japanese aesthetics with the latest modern technology, an idea he referred to as "Japanese Modern."

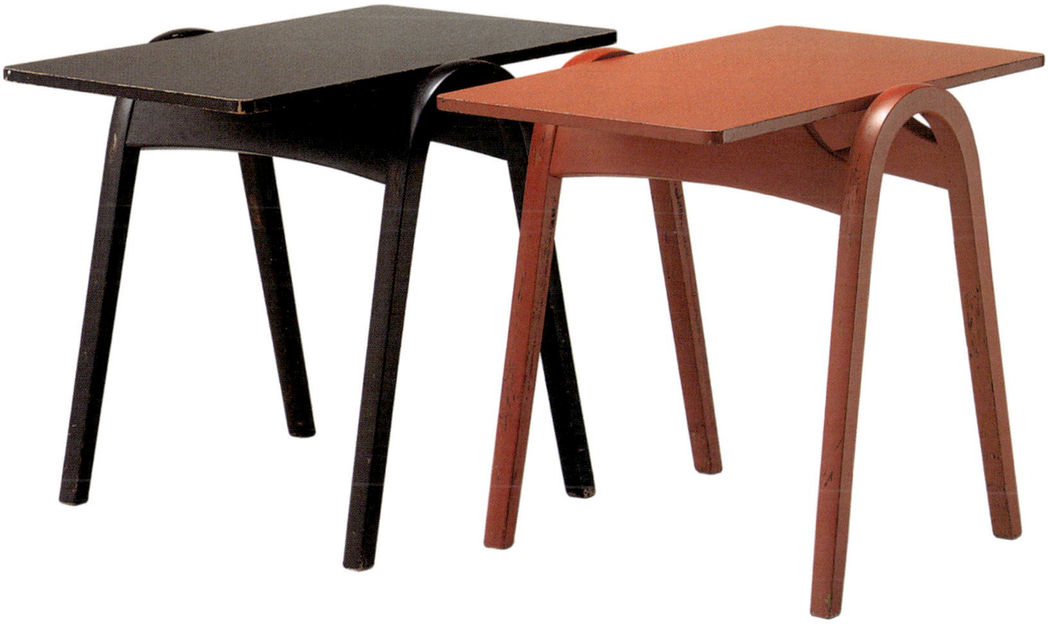

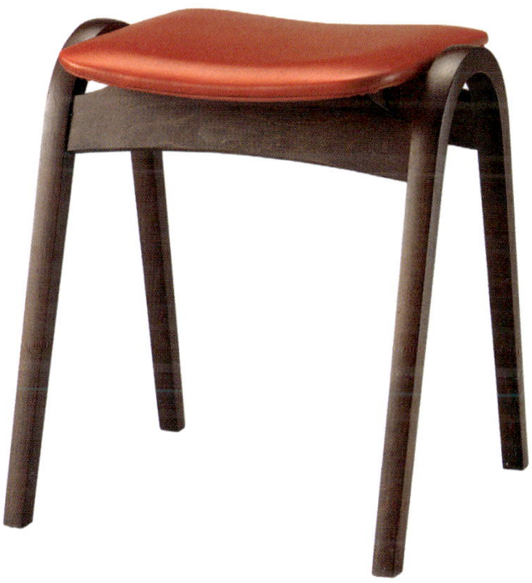

Top: Kenmochi Design Associates, Tea Tables, beech, 1958, 16 x 23 x 18 in. Bottom: Kenmochi Design Associates, Stacking Stool, beech and vinyl, 1959, 17 5/8 x 15 5/8 x 13 5/8 in., manufactured by Akita Mokko Co., Ltd. While at the Industrial Arts Research Institute, Kenmochi experimented with bentwood techniques for mass-produced furniture. This lightweight and easily stackable stool conserved space in the modest-scaled domestic interiors of the 1950s and 1960s. The stool was first shown at the Ginza Matsuya Department Store *Apartment Furniture* exhibition in Tokyo in 1958, a time when postwar Japan was experiencing a boom in urban apartment building. Popular since it was first distributed in 1959, it is still in production today.

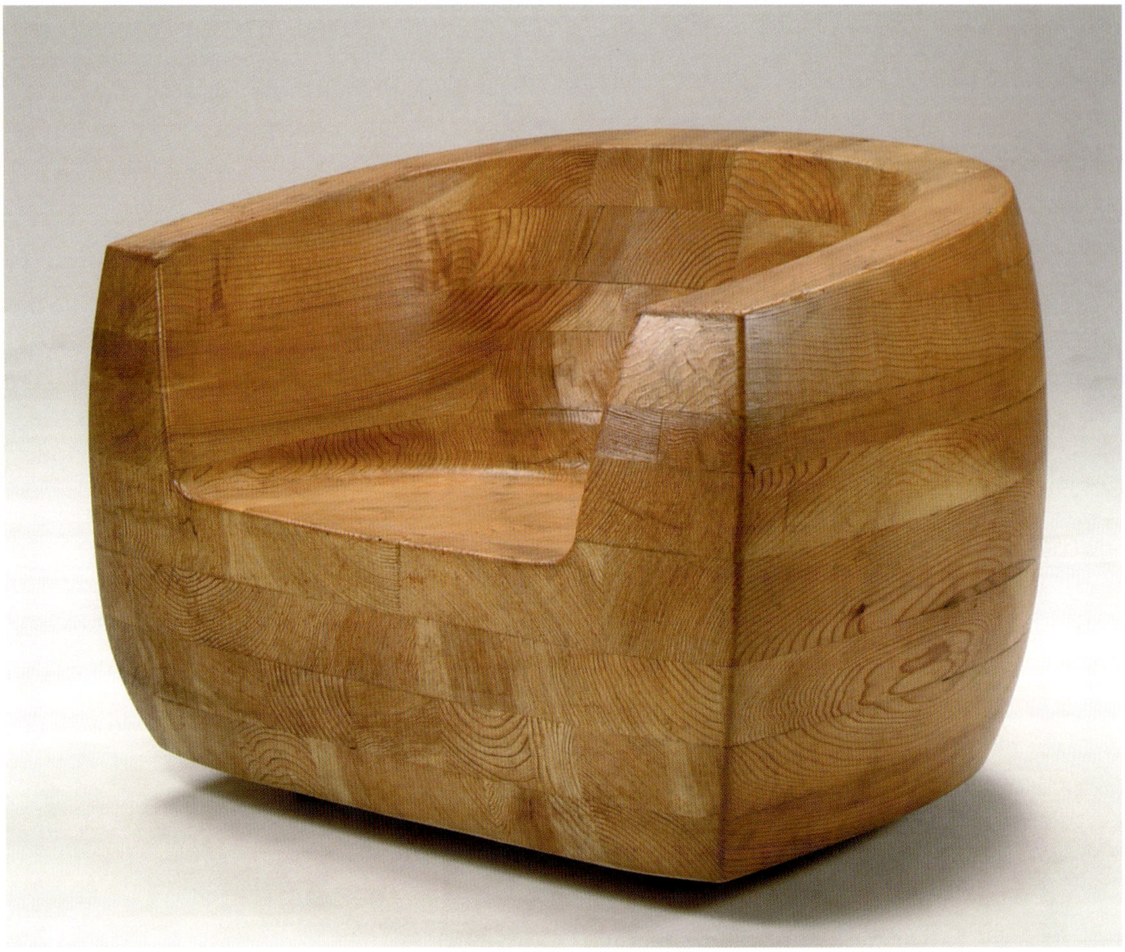

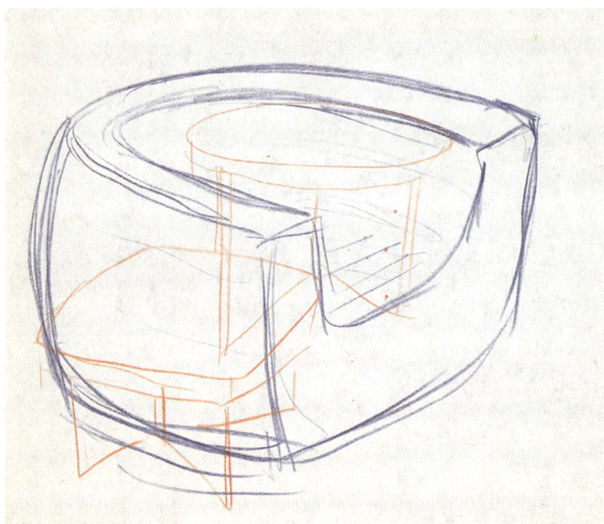

Top: Kenmochi Design Associates, Kashiwado Chair, cedar, 1961, 25 x 30 9/16 x 27 15/16 in., manufactured by Tendo Co., Ltd. Originally designed for the Atami Garden Hotel in Shizuoka Prefecture, famous for its hot spring resorts, this chair was one of the first joint projects between Kenmochi Design Associates and Tendo Co., Ltd., which began to employ outside designers on royalty contracts from the 1950s. The chair is built with blocks of Japanese cedar in a unique U-shaped profile, with the grain arranged in a decorative pattern. The stout, imposing shape of this armchair inspired its name, Kashiwado, after a famous Sumo wrestler of the period. Bottom: Isamu Kenmochi, Kashiwado Chair, blue pencil on tracing paper, ca. 1960, 8 x 9 1/2 in.

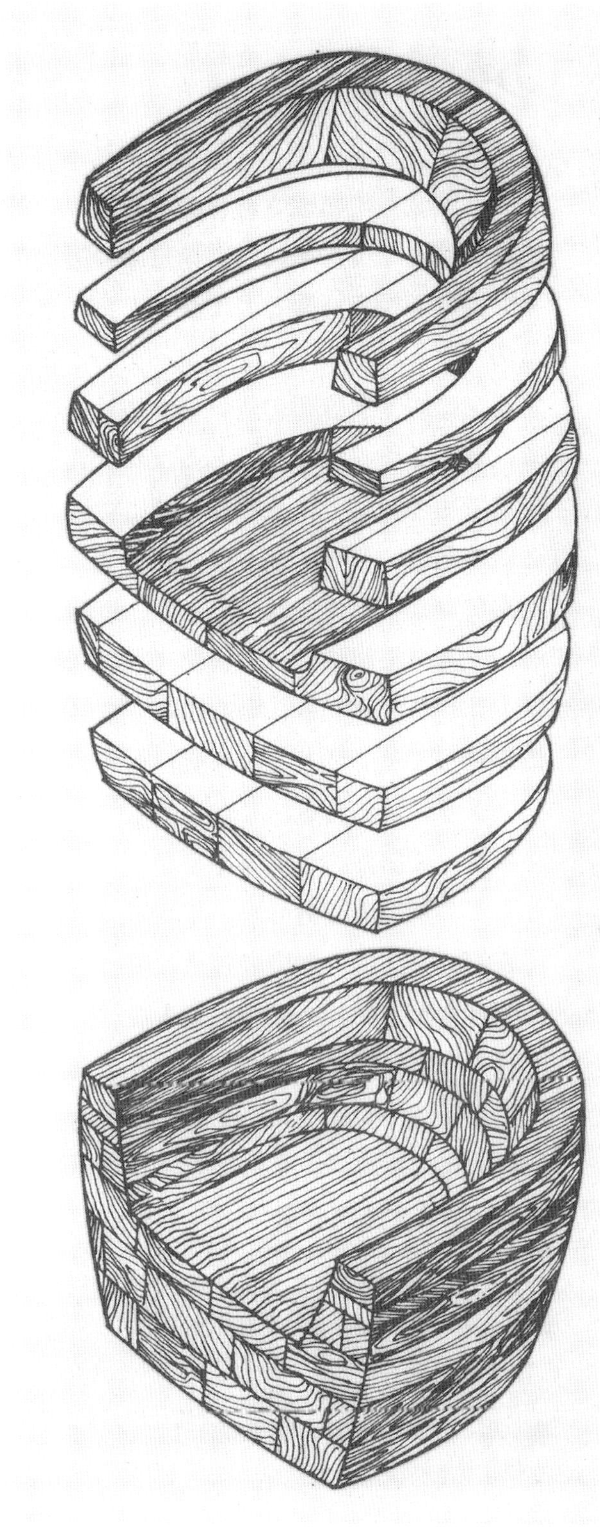

Kenmochi Design Associates, Kashiwado Chair, technical drawing, ca. 1961.

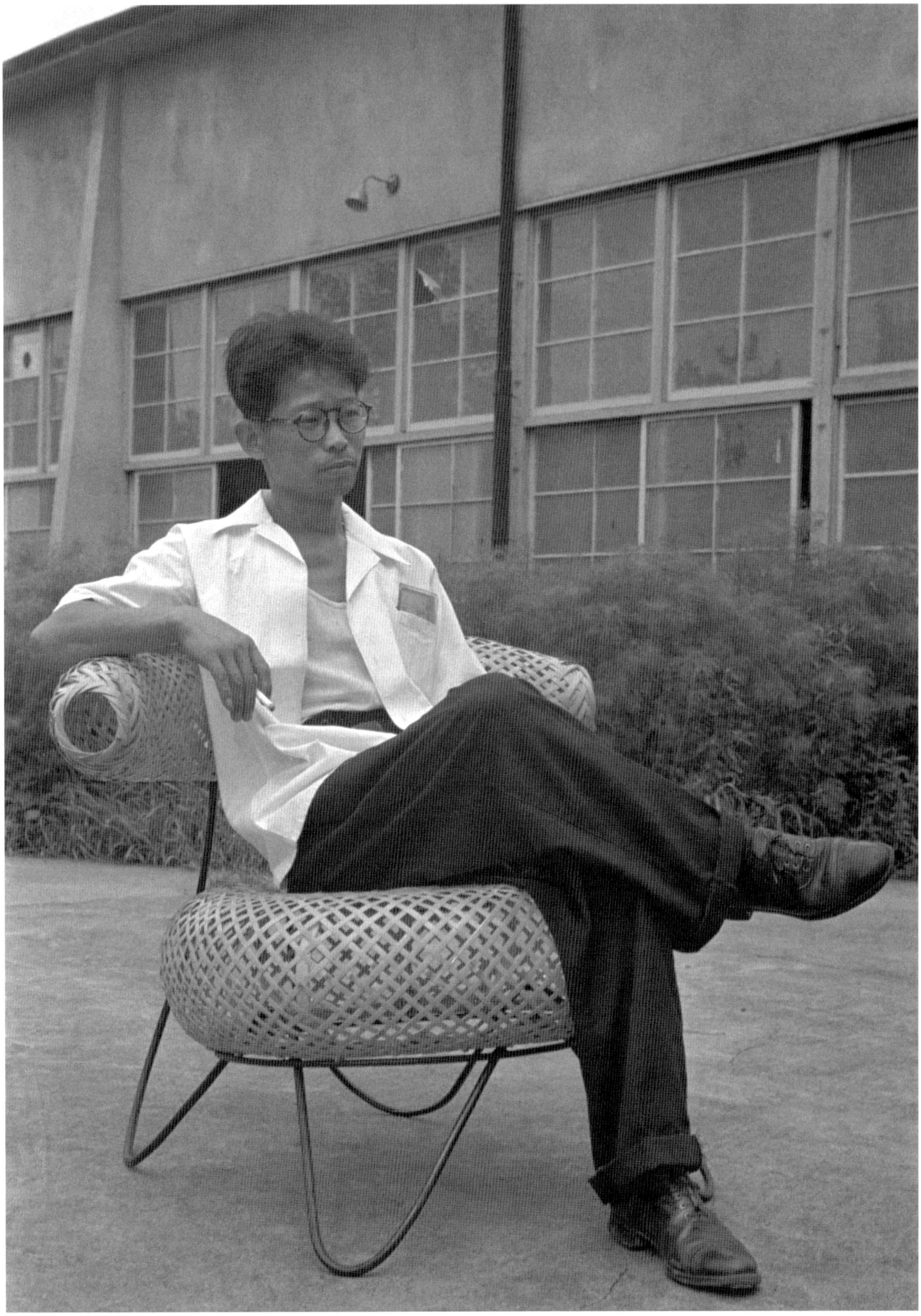

Photograph by Isamu Noguchi of Harumitsu Inoue, IARI craftsman, seated on the Bamboo Basket Chair designed by Isamu Kenmochi and Isamu Noguchi, 1950.

ISAMU KENMOCHI AND JAPANESE MODERN DESIGN

TETSUO MATSUMOTO

In 1952 Isamu Kenmochi set sail from Yokohama port, aboard a passenger freighter, for his research trip to study the American economy, industry, and its latest designs. The budget was limited, yet he managed to gather extensive information, at times following suggestions given by Isamu Noguchi. He traversed the continent eastward by bus. The journey lasted seven months, from April 30 until November 29. Only in April of that year, Kenmochi assumed the position of Director of Design at the new Industrial Art Institute (IAI), formerly known as the Industrial Arts Research Institute, which once served as a temporary studio for Isamu Noguchi as he worked on his *Shin Banraisha* project.

America at first-hand was an eye-opening experience for Kenmochi, staggering especially in comparison with the situation in Japan. Throughout his journey, he transmitted an "American Newsletter" to the IAI. In it he reported on his meetings with figures such as the Eameses, Nelson, Mies, Girard, Rudofsky, Robsjohn-Gibbings, Kaufmann, Wachsmann, Breuer, and Saarinen, and wrote about his visits to the Boston Museum and the *Good Design* exhibition, and about popular lifestyle in general.

In particular, the numerous examples of "Japanese Beauty" brought by Americans into their daily lives astounded him. Well-designed bamboo crafts and lacquer-finished bowls and trays were displayed and sold as modern design, while some buyers were not even aware that they were Japanese. He saw some Japanese crafts shown at the *Good Design* exhibition, and was saddened a little when he did not see any credit given to designers' names. What started as simple learning and absorption would gradually grow into the cultural comparison of Japan and America, and finally to a state of self-reflection. He saw a gap too large between Japan and America, where the originality of designs defined their manufacturers and supported capital markets.

Back in Japan, Kenmochi discussed the prevalence of Good Design in American daily life and of Japanese design found there. "Modern style in American architecture and crafts is surprisingly not American," he wrote. "The mood that fills them is modern but they have classical oriental taste. Orientalism is one of the trends in Europe, but I never expected Japanese style to be so popular in America." To him it was a surprise that Modernism in America was similar to Japanese traditional crafts. He conveyed the words of an importer that he "wished to import not antiques but modern designs from Japan." Or that "the works of Paul McCobb appear to be something Japanese designers should be making. Robsjohn-Gibbings' library was full of old and rare books about Japanese architecture and industrial arts." Also quoted by him are modernists Edgar Kaufmann, Jr.—"This year's American modern has its origin in strict medieval style"—and Bernard Rudofsky: "There is no longer a distinction between the East and the West. As there never should have been . . ."

Kenmochi's own words express his aspiration for the future of postwar Japan, reflecting the seven months abroad that enabled him to view Japan objectively and his double duty as a designer and an official at the Ministry of Commerce and Industry: "We seem to have lost ourselves in following the events of abroad, and are about to lose the strength we always had. Our efforts to observe and keep up with the cultures abroad have at some point changed into mimicry. Especially when there is admiration for them, have we not willingly lost the

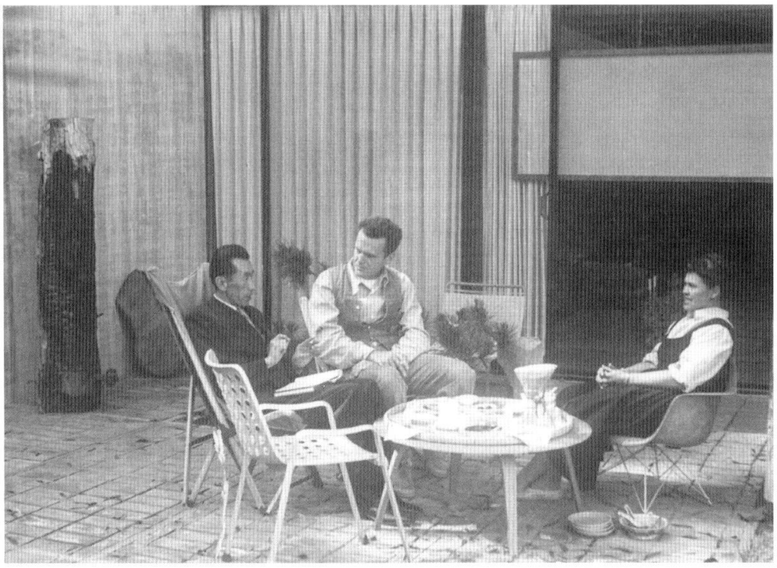

Isamu Kenmochi and Charles and Ray Eames at the Eames residence in Pacific Palisades, California, 1952.

values that earned such respect? For modern technology, there is still much more to be learned from abroad. But in terms of creation, when we execute our creativity with thorough deliberation, there is no need to be concerned with the fashionable aspect of the foreign works."

These thoughts, considering the country's social circumstance at the time, were rather emotional, yet they were as unique as he was, and he would later prove them with his actions.

Kenmochi's first report on America was made at a time when American designs were introduced to influence Japanese designers in various fields, including the IAI. While Kenmochi was in America, a country flying high in its postwar prosperity, and where he came across housewares from Scandinavia—especially the so-called Swedish Modern—he must have thought it possible for Japan to do the same, to make its way into the American conservative trend. Of course, he was aware of the high standard of design in Scandinavian lifestyle, but as a technocrat at the Ministry of Commerce and Industry, he felt obligated to improve export gains and earn dollars. His reports on America earned IAI a special research grant for three consecutive years from the Ministry of Finance.

According to Kenmochi, Japanese Modern was Japanese Good Design. He was aware of the gap between his wish to export Japanese Good Design, and the reality that Japanese daily life was still dragging its feet since the war. Perhaps it was his priority as a public official that sent him down the path of the "Japonica Controversy." The year was 1953, when Japanese architects had started their own "Tradition Controversy," and those who were deeply immersed in postwar American culture were beginning to feel an urgency to find their own Japanese modernism.

Deep down in his consciousness, Kenmochi knew "Japan as Tradition," which may have been close to what Yuichiro Kojiro called "the vernacular (the red)," but he had no self-awareness of it. As a result, only "the nobility (the white)" was pushed forward and materialized.

His advocacy of Japanese Modern earned its legitimacy as a result of combining various factors: the Japanese domestic situation, his serious nature, his intuitive aesthetic sense and his strong will. His curiosity was relentless, and was directed toward all objects and events. Whenever he picked his subject matter, his research was broad yet intense. He showed the same interest towards humans as well. The color slides he took demonstrate the way he was always closing in on his subjects. Perhaps it was his determination to discover the true nature of all things.

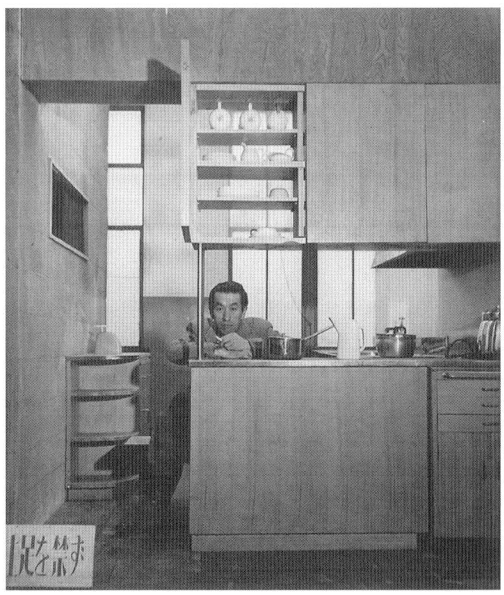

Photographer unknown, portrait of Isamu Kenmochi, ca. 1950s.

During the years between 1952 and 1955, Kenmochi ruminated over the future of Japanese design and designers. As an official of the Ministry, his comments and proposals only enlarged the separation between Japanese society and its developing democracy. At the same time, as he became acquainted with foreign designers and architects and their status abroad, he felt the need to let society, and the designers themselves, know the raison d'être of their profession. The founding of JIDA (Japan Industrial Designers Association) was part of such an effort, although the problem soon reverted to the issue of the system and the individual.

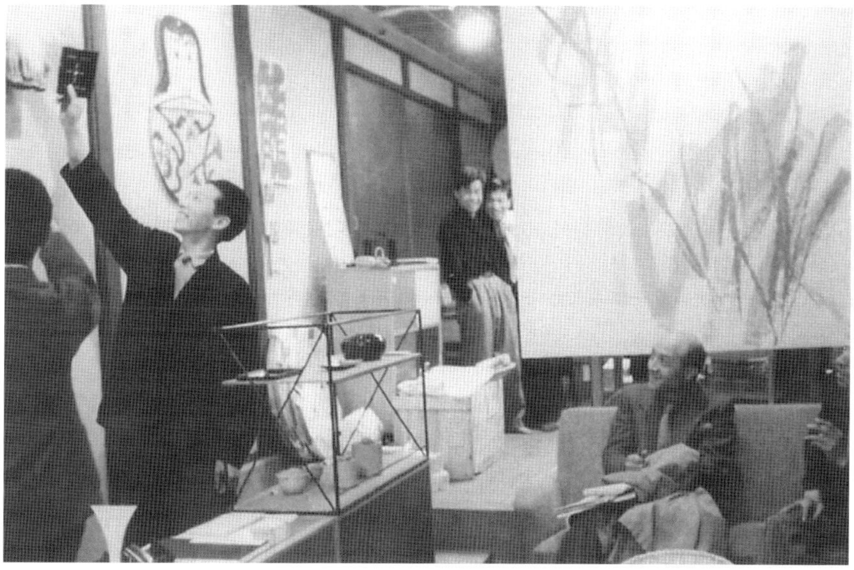

Isamu Noguchi (right) at Kenmochi's Living Art Showroom, ca. 1956–1958.

Interior designers at the time had not earned their independence in society. From the beginning Kenmochi felt that they deserved a status equal to architects, most likely because he knew many architects and artists as friends. Upon setting a new goal to realize his ambitions—to produce Japanese Good Design in the realistic business environment—he resigned from the Ministry in June of 1955 and established Isamu Kenmochi Design Associates.

Back then, all designers were assumed to be fashion designers, hence his frustration. A new profession can become established and reliable only when the conditions and timing demanded by the society are matched by the technical standard and quantity supplied by the profession. So the opening of the Kenmochi office may have been premature, but I must also add that his work as a designer was always a mixture of two issues: of design and of the designer's profession. This was illustrated when he said, "Japanese furniture is too commercialized. Architecture is large and furniture is small. Architecture is academically systematized, and yet furniture does not call for much specialty. But should furniture be treated as mere commodity? To this I must strongly disagree. It is because furniture exists within architecture. As long as that is the case, I believe that furniture must always be at the same cultural level as architecture, as high as today's architectural standard brings itself to be." He was true to this principle, and demanded a high design philosophy and technical standard of manufacturers as well. By associating only with the manufacturers who responded to this demand, he collaborated in product development. Through production of good design, it was his belief that designers and manufacturers can contribute to society, and bring success to companies.

In summer of 1970 Kenmochi spoke on the radio as follows: "The Japanese spend money on food and clothing but don't have a standard of spending for their dwellings. It is not good to simply go by the price without thinking how many years it will last. Good furniture may be expensive, but if you plan to use it carefully through generations you cannot buy a cheap item. Without this understanding the Japanese living standard will never reach the level of America and Europe. Spend money on good things that are worth it. A hardworking designer alone cannot create good design. Arrogant manufacturers and merchants will not make a good design. Pretentious consumers will not make good design. Designers with real skills should assert themselves as creators. Manufacturers should produce at good quality and set a fair price. Consumers should insist on getting the right product. When the balance of the three is finally met, a good product is made and then society can finally enjoy a healthy growth in the culture of lifestyle. But in today's Japan, this triad is off balance." The earnest tone reveals his serious nature, and his usual optimism is now forcing him to preach in a grave manner with a touch of somberness. Every time I am reminded of this passionate speech, I sense the end of an era. It was only a year later that he committed suicide. Sometimes I wish I could ask him whether he would consider himself lucky for not having witnessed the design world since the 1970s.

Kenmochi experienced three different external viewpoints of Japan: Bruno Taut in 1933, Charlotte Perriand in 1940, and Isamu Noguchi in 1950. The first two were architects and designers and elite intellectuals from Europe. But Kenmochi felt an affinity with Isamu Noguchi and his Japanese genes, and with his philosophy, ideology, and sensibility, and found common traits in their views of Japan.

I remember when Noguchi visited our office in Aoyama in about 1957. He pointed to a vertical lamp in the office and said, "Kenmochi-san, this looks like mine." The lamp was designed for a project in a Shiseido Beauty Salon in 1957, soon after the office had opened. I remember Kenmochi explaining to Noguchi, "It is derived from a lesson at the Bauhaus in which you create a three-dimensional geometric form out of a single piece of paper. So it is different from yours, in shape and material."

It is true, however, that Kenmochi's lifelong theme of Japanese Modern would not have been born were it not for his meeting with Isamu Noguchi in 1950.

Translated from the Japanese by Eiko Sakai and Wayne and Noriko Peet.

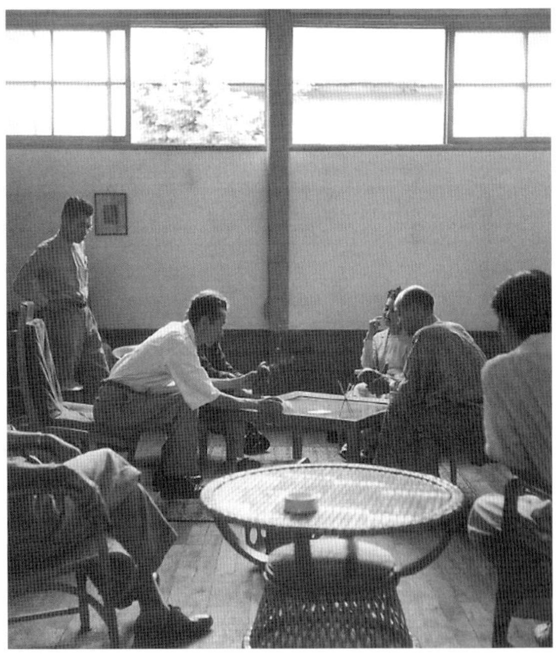

Isamu Kenmochi (seated, center left) and Isamu Noguchi (seated, center right) at the Industrial Arts Research Institute, 1950.
Opposite: Isamu Kenmochi, 1955.

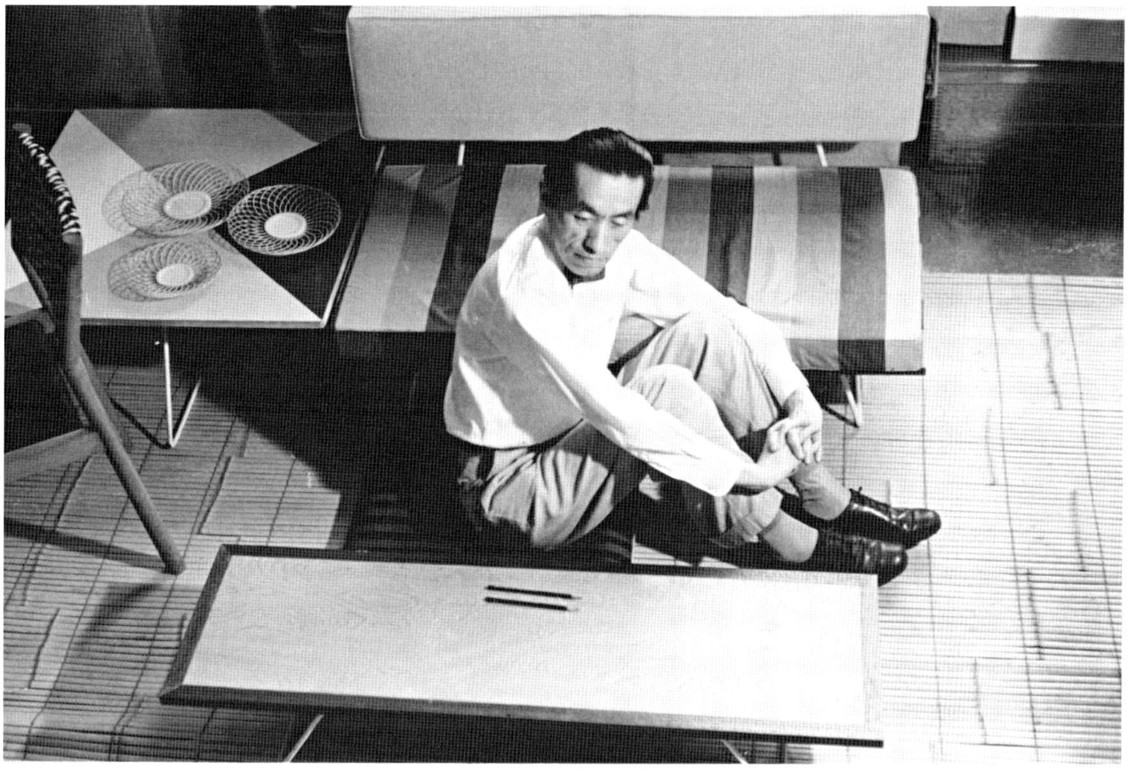

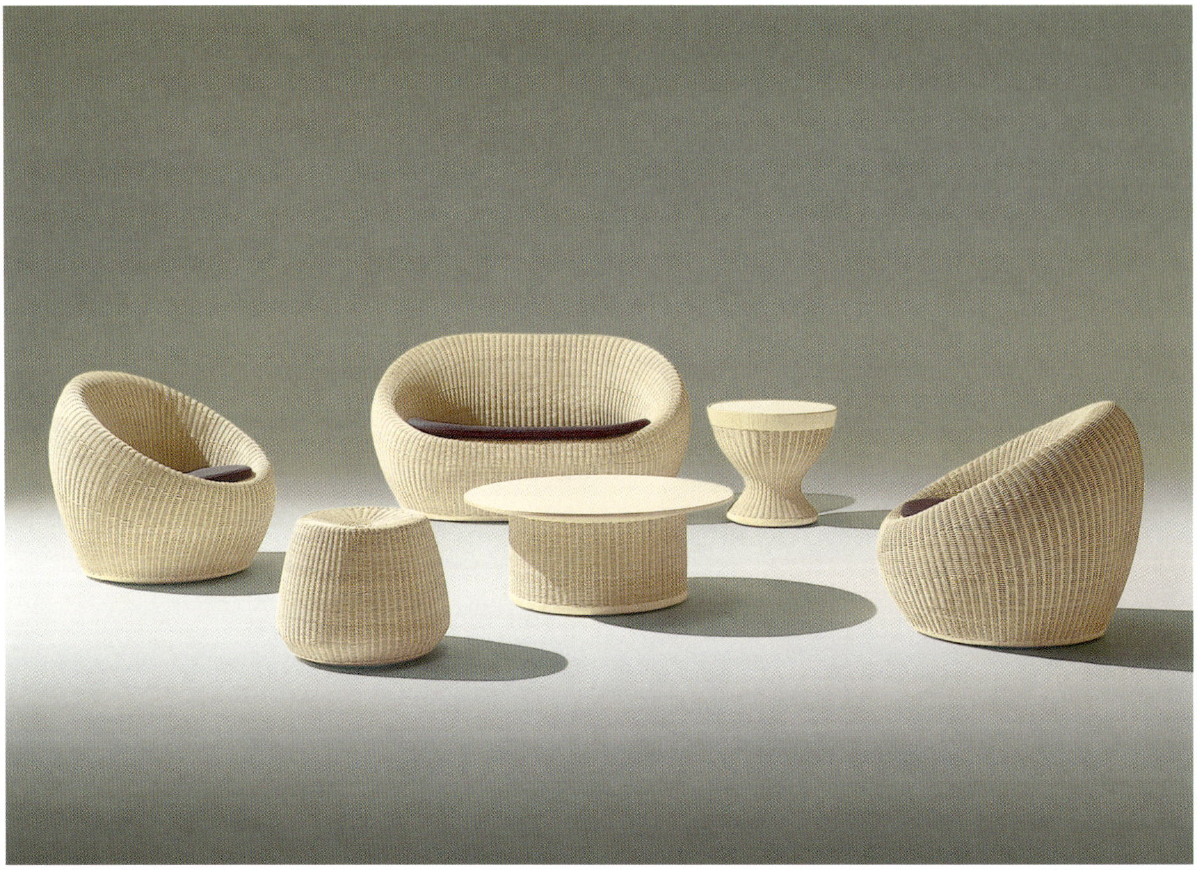

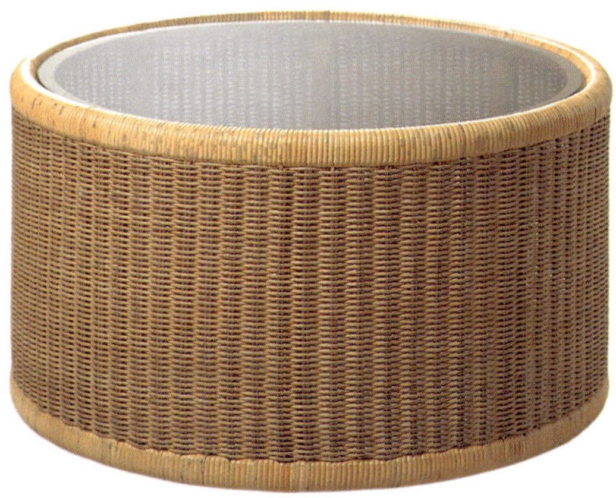

Top, clockwise from left: Kenmochi Design Associates, Rattan Round Chair C-3150, 1960, 27 9/16 x 31 15/16 x 27 9/16 in.; Settee, 1967, 27 9/16 x 51 3/4 x 30 11/16 in.; Table T-3030, ca. 1960s, height: 14 in., diameter: 35 in.; Rattan Round Chair; Table T-3010, 1965, height: 18 in., diameter: 21 in.; Stool S-3030, ca. 1960s, height: 15 in., diameter: 17 in. Bottom: Kenmochi Design Associates, Table, rattan and glass, 1964, height: 12 5/8 in., diameter: 23 5/8 in.

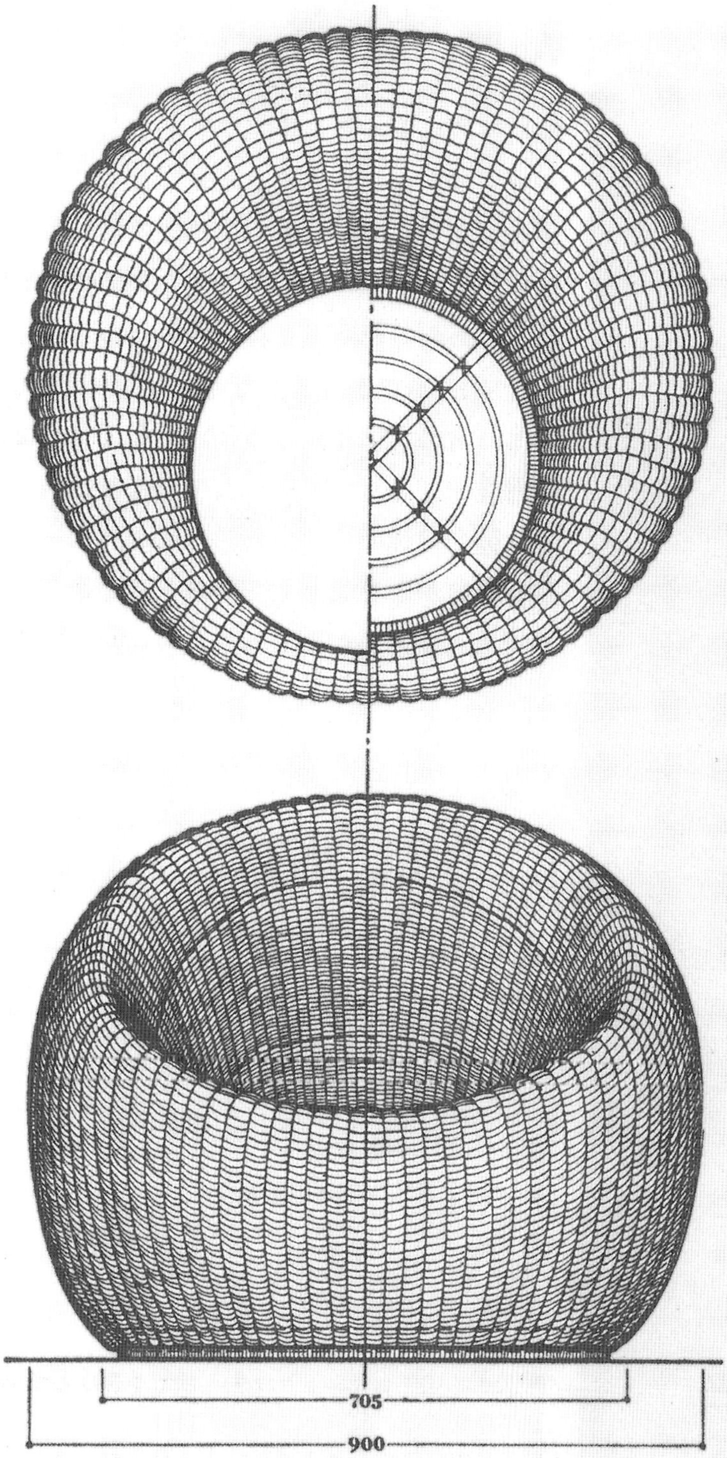

Kenmochi Design Associates, Rattan Round Chair, technical drawing, 1959–1960. Selected for inclusion in the permanent collection of The Museum of Modern Art, New York, in 1964, the Rattan Round Chair made Kenmochi's international reputation as a furniture designer. The rattan chairs embody Kenmochi's design principle of the perfect match of the human and the natural. The Rattan Round Chair was initially handmade by Yamakawa Rattan in 1959, and variations of the design were developed throughout the 1960s. It is still in production today, made one by one, and distributed by Y.M.K. Co., Ltd., in Japan.

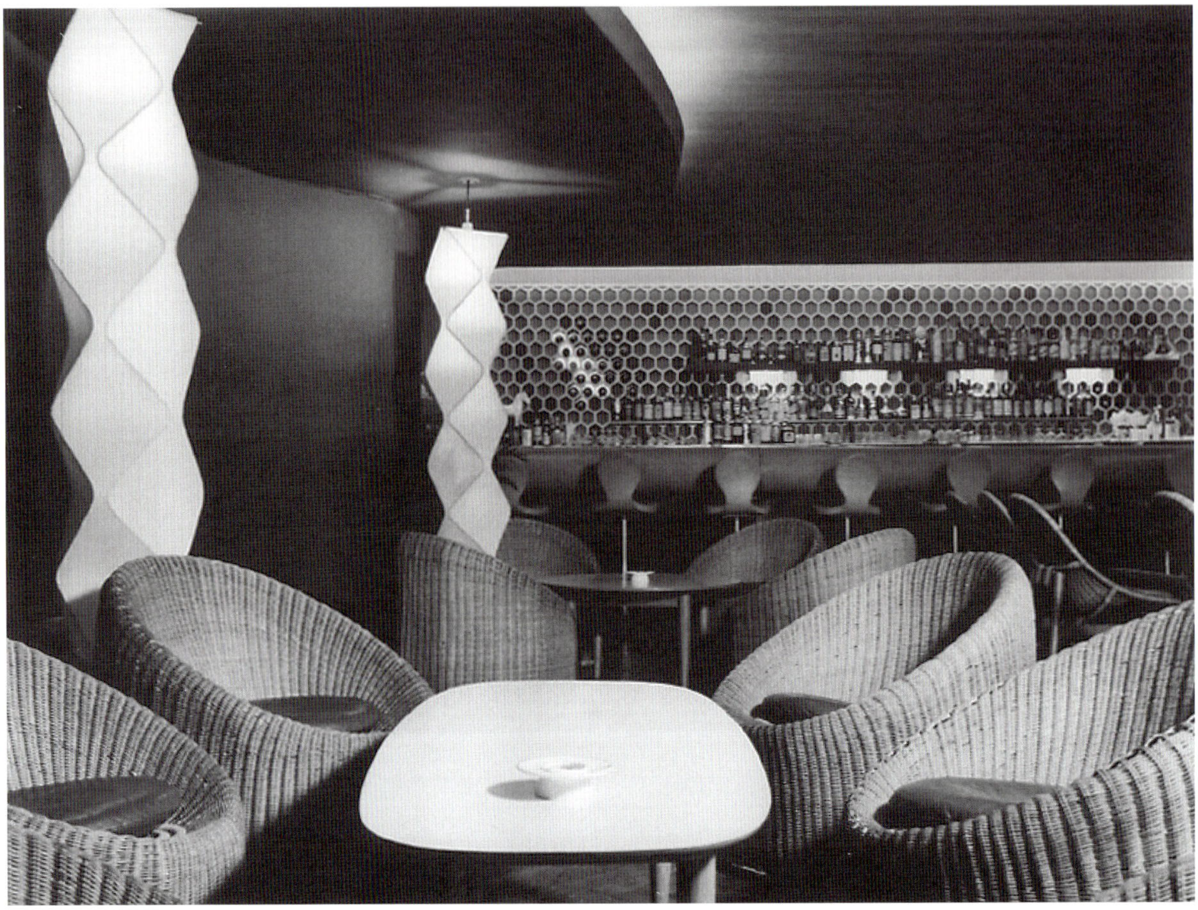

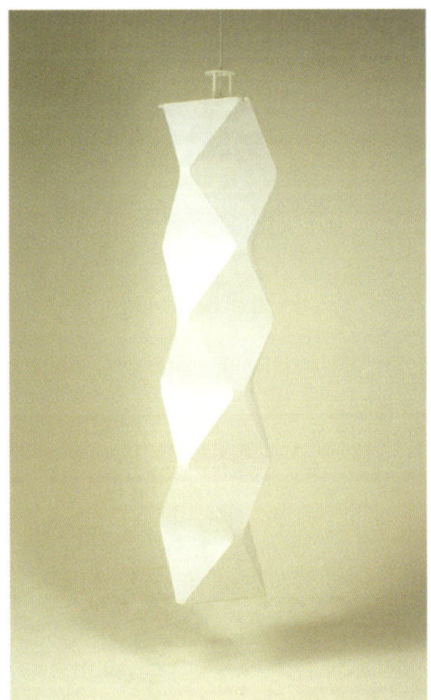

Top: Kenmochi Design Associates, interior, main bar at the Hotel New Japan, 1960. The Hotel New Japan in Tokyo was planned by Takeo Sato Architecture and Associates. It was the first large international hotel built after World War II. Kenmochi Design Associates was commissioned to design the hotel's interiors. The hotel was intended to combine Japanese and European style, and before designing the interiors, Kenmochi traveled to the United States and Europe to study several first-class hotel interiors, including Hilton hotels in Beverly Hills, Berlin, and Istanbul. Kenmochi was inspired by the way in which these hotels integrated local motifs with a modern vernacular. Bottom: Kenmochi Design Associates, Hanging Lamp, iron and fiberglass, 1960, height: 55 5/16 in., diameter: 10 3/4 in. The Hanging Lamp was first developed in 1957 for the interior of the Shiseido Beauty Salon in Ginza, Tokyo.

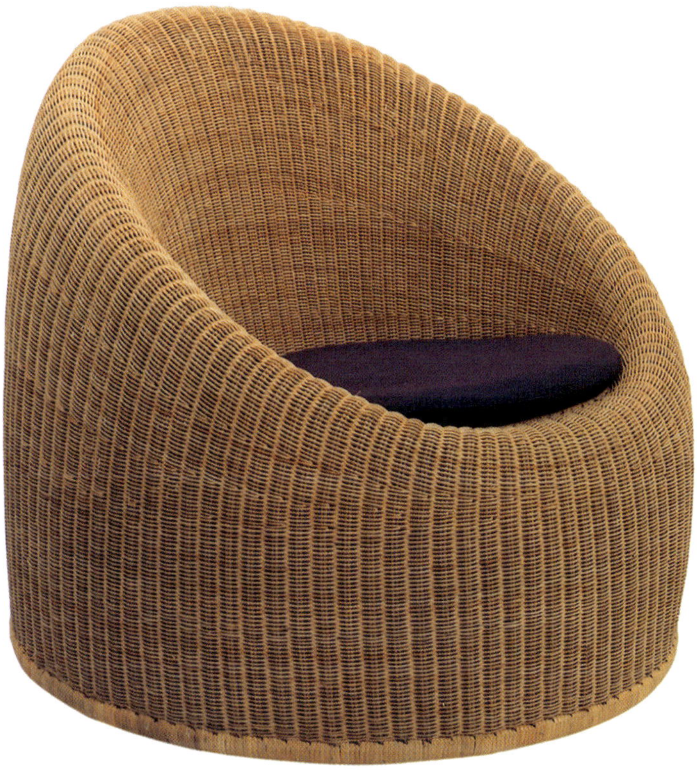

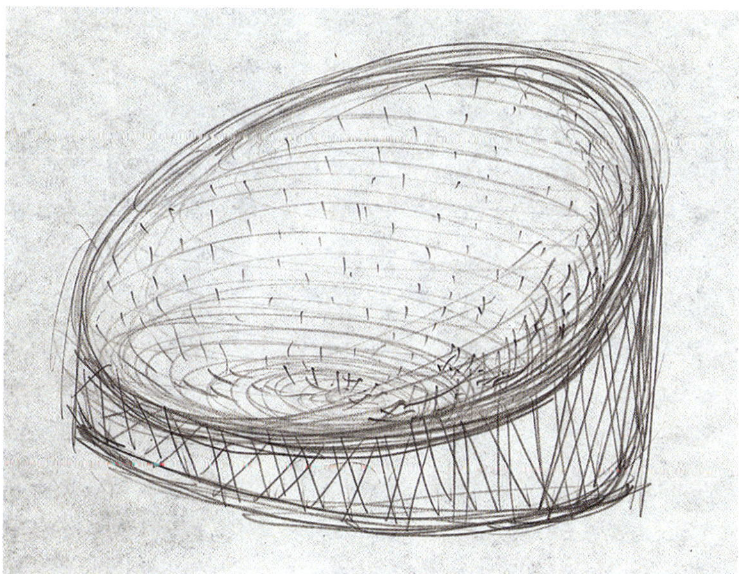

Top: Kenmochi Design Associates, Rattan Chair, 1959, 27 9/16 x 35 7/16 x 34 11/16 in., manufactured by Yamakawa Rattan (current production by Y.M.K. Co., Ltd.). Bottom: Isamu Kenmochi, Lounge Stool, pencil on paper, ca. 1959, 10 11/16 x 16 1/8 in.

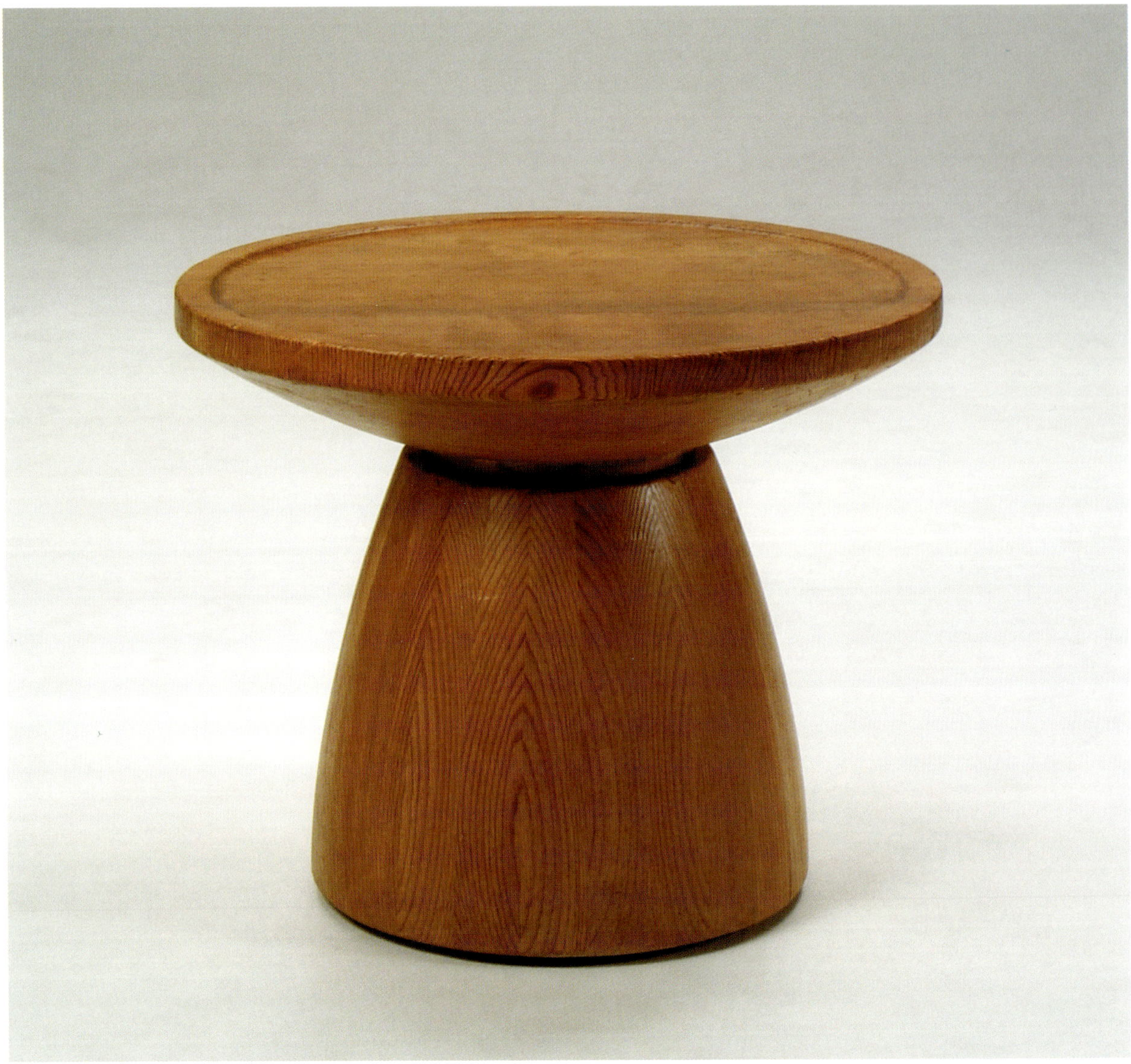

Kenmochi Design Associates, Table, cedar, 1961, 17 7/16 x 20 15/16 in., manufactured by Tendo Co., Ltd.

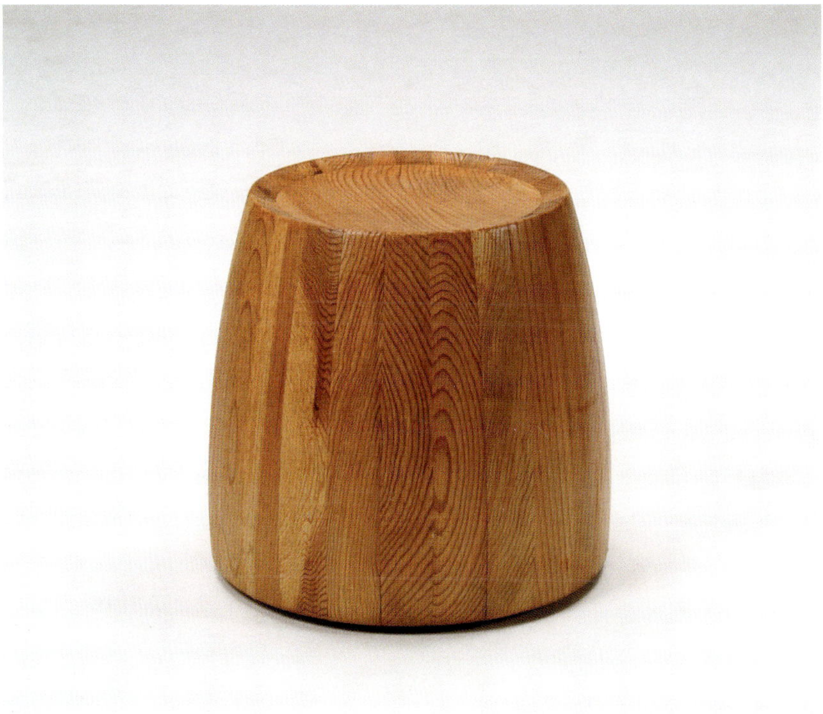

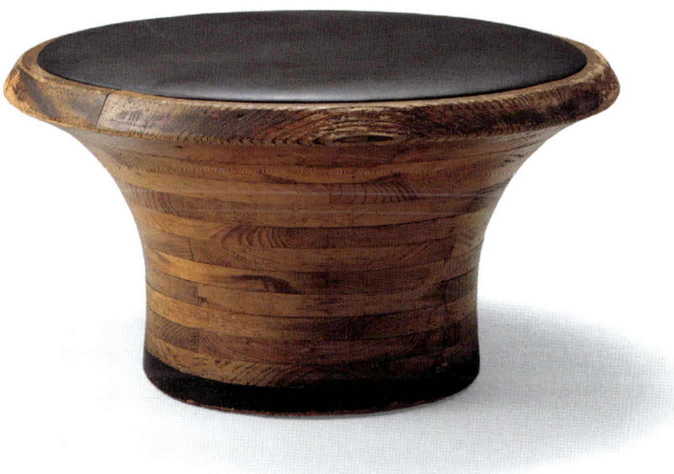

Top: Kenmochi Design Associates, Stool, cedar, 1961, height: 13 9/16 in., diameter: 14 3/16 in., manufactured by Tendo Co., Ltd. The matching Kenmochi Table (opposite) and Stool were originally designed for the Atami Garden Hotel in Shizuoka Prefecture, Japan. Bottom: Kenmochi Design Associates, Circular Bench, cedar and vinyl, 1965, 17 7/16 x 33 1/2 in., manufactured by Tendo Co., Ltd. Designed for Kagawa Prefecture Gymnasium, Takamatsu, Japan (Kenzo Tange, architect).

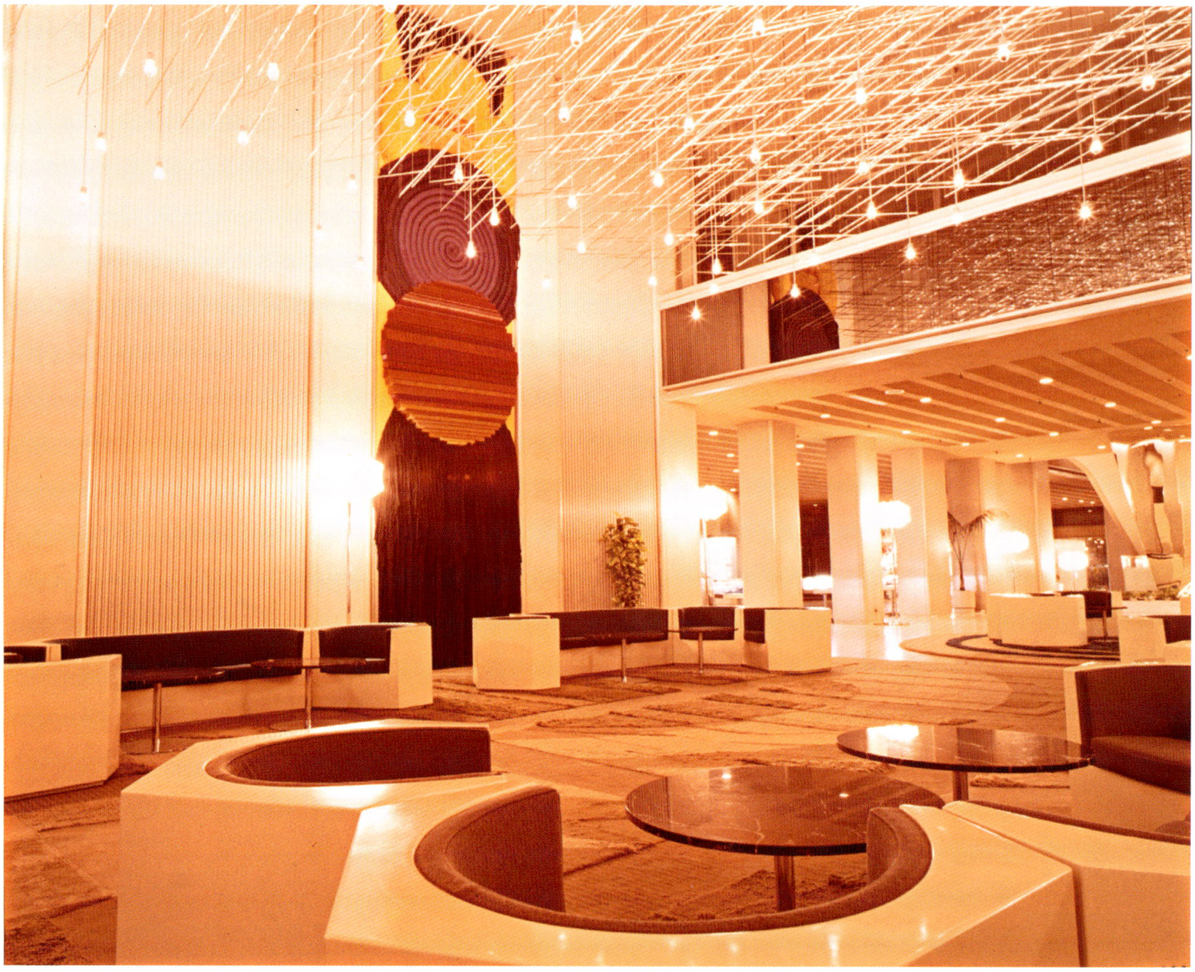

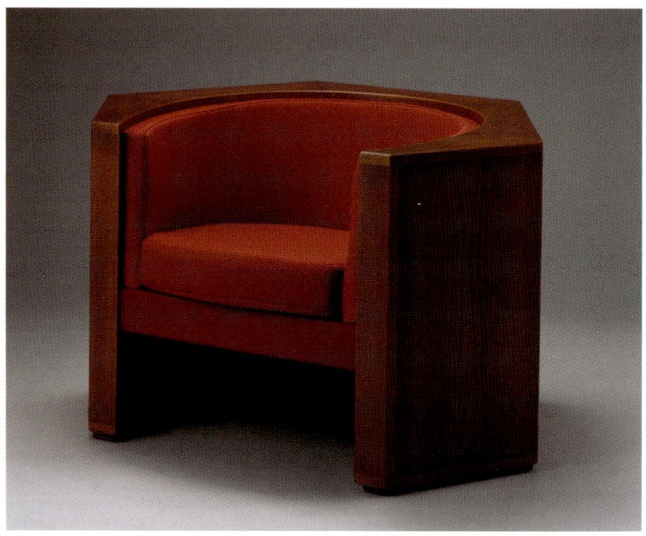

Top: Kenmochi Design Associates, Entrance Lobby, Keio Plaza Hotel, 1971. Bottom: Kenmochi Design Associates, Easy Chair F, 1966, 24 7/16 x 39 13/16 x 29 1/8 in., manufactured by Tendo Co., Ltd. Opposite: Kenmochi Design Associates, Main Hall, Kokusai Kaikan (Kyoto International Conference Center), 1966.

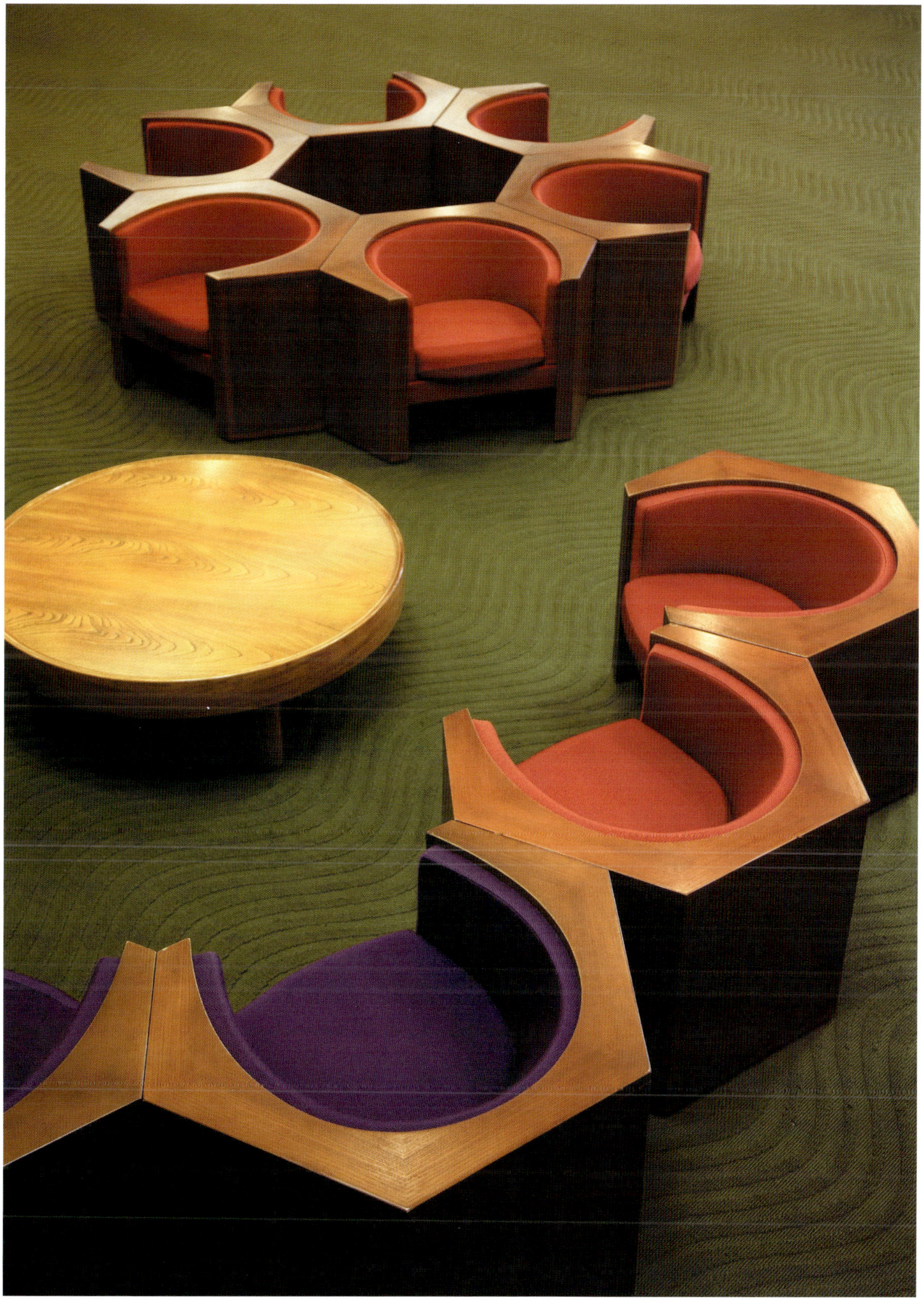

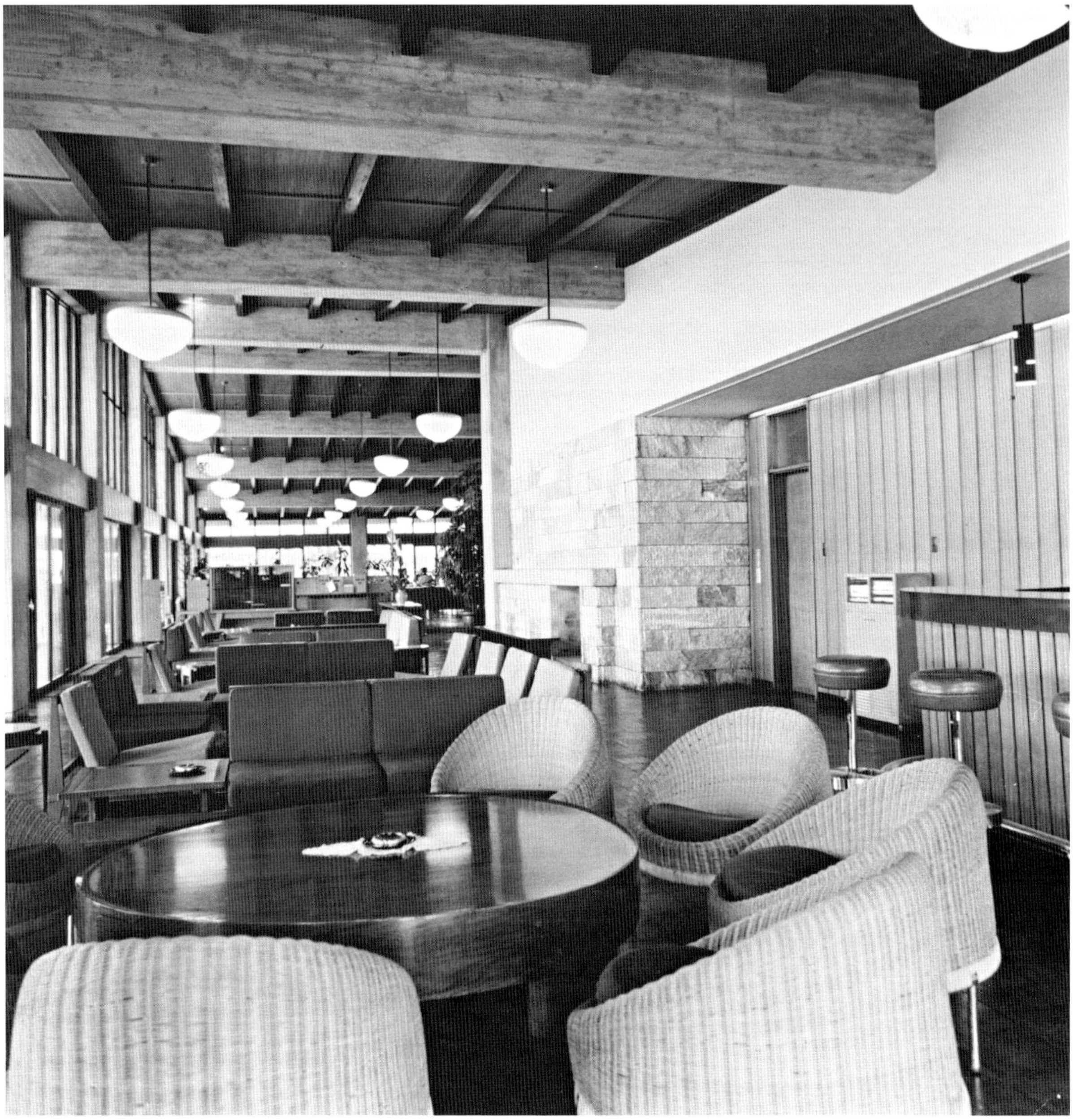

Kenmochi Design Associates, Lounge Table and Chairs, Ohtone Country Club, Ibaraki Prefecture, Japan, 1960.

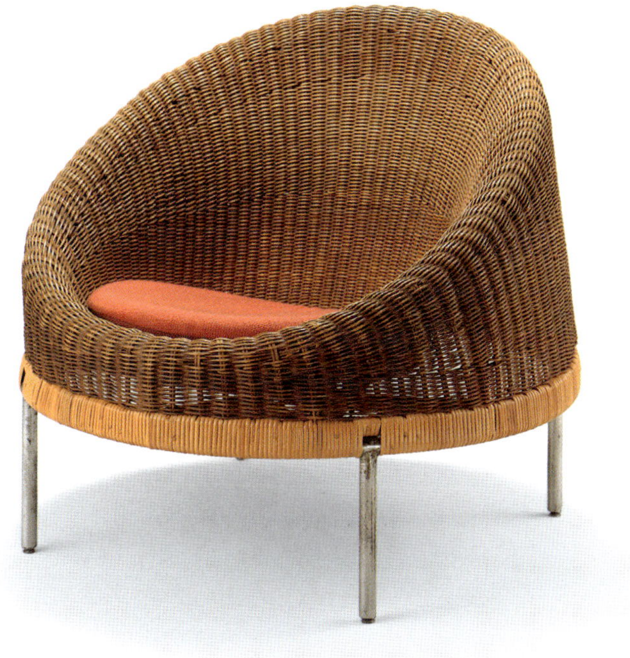

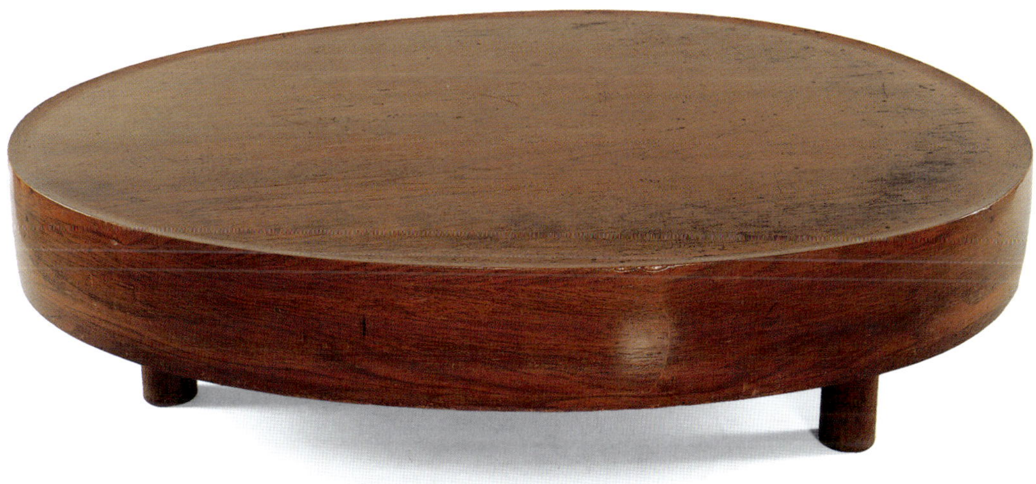

Top: Kenmochi Design Associates, Lounge Chair for Ohtone Country Club, rattan and iron, 1960, 27 3/4 x 27 9/16 x 32 5/16 in., manufactured by Yamakawa Rattan. Bottom: Kenmochi Design Associates, Lounge Table for Ohtone Country Club, oak, 1960, 17 11/32 x 64 9/16 x 56 23/32 in., manufactured by Tendo Co., Ltd. Japan's postwar economic recovery brought new leisure activities to an increasingly affluent public. The expansion of facilities such as hotels, golf clubs, and stadiums created a demand for modern interior designers and architects. Kenmochi was commissioned to design furniture, including the lounge chair and table above, for the Ohtone Country Club in Ibaraki Prefecture. The table's short legs evoke a traditional Japanese living environment but the boldness of its thick oval shape was a modern innovation meant to establish a strong presence in the club's open interior space.

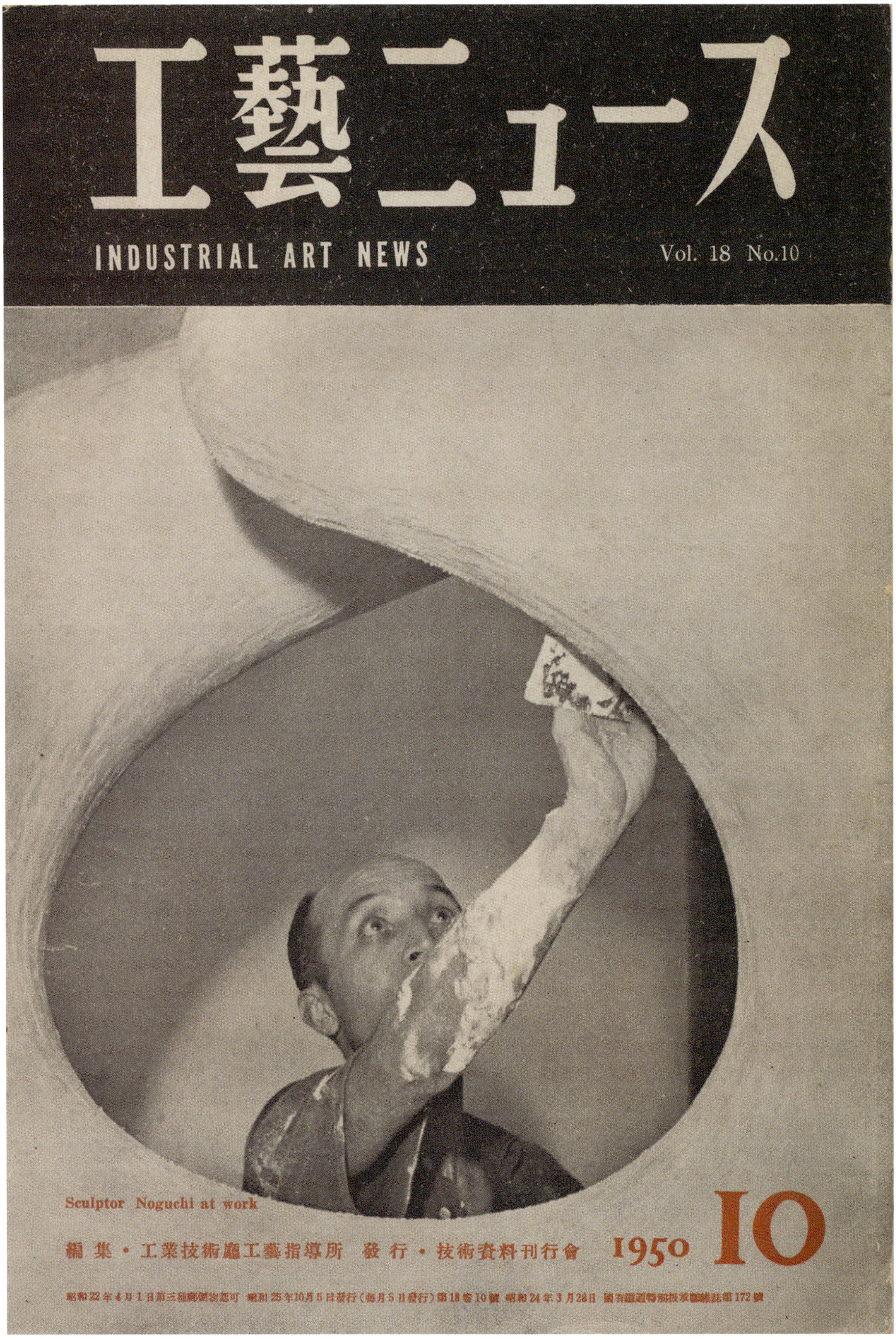
工藝ニュース
INDUSTRIAL ART NEWS
Vol. 18 No.10
Sculptor Noguchi at work
編集・工業技術廳工藝指導所 發行・技術資料刊行會
1950 10
昭和22年4月1日第三種郵便物認可 昭和25年10月5日發行（毎月5日發行）第18卷10號 昭和24年3月28日 國有鐵道特別扱承認雜誌第172號

MODERN TRENDS IN INTERIOR FURNISHINGS

Excerpt from a lecture at the Tokyo National Museum, June 20, 1950.
From *Industrial Art News*, Vol. 18, No. 10, October 1950, pp. 18–19

ISAMU NOGUCHI

American Trends

Furniture design in America today seems to focus on exaggerated practicality. This is not a good thing. On the other hand, extreme décor camouflages function. I am not fond of such "decorations" either.

A window need not be merely a window. If you think of a window as a wall of light, then you can eliminate the conventional idea of a window and now it becomes a wall that is also a source of light.

The new house in America is shedding its unnecessary functions one by one. If its functions are too visible, they will also be eliminated. This will bring forth a simplification of the forms of all interior and daily objects. An object needs to find its space. If it is a functional object, it just needs to be beautiful. All useless decorations must be removed, and that is one aspect of our new furniture. The new furniture in America is also losing its traditions. However, tradition in furniture has a practical psychological function. For example, a primitive flame has a psychological effect, which is why we yearn to have a fireplace in a house or a cottage. A fireplace is a source of warmth and light, satisfying a psychological demand of human nature. In contemporary rooms, standing lamps have been used to answer this demand, and such sources of warmth and light have always helped to provide a perception of safety.

We can also say the same for a rug, which creates a sense of psychological trust over a particular area. An older house may have many private rooms, but in recent houses, there are fewer walls, thus fewer rooms. This will take some courage, for Americans are rather sensitive to openness, which may fit the Japanese lifestyle more easily. I can say that this openness is being forced upon Americans today, creating a pressure point for young American architects of this trend.

Advocates of this trend toward openness would do anything to acquire the modern style with roots in Japanese culture—a chair so low it resembles a futon, a soft floor like a tatami mat, or walls as softly luminous as shoji screens. But the chairs of this new trend are out of scale with the body types of average Americans. This imbalance between body and chair seems underscored by the fact that we never see a human figure included in the interior shots reproduced in architecture magazines. Low and skinny furniture is the new rule. This type of furniture is more fitting for Japanese bodies, and the Japanese seem more suited to the modern lifestyle such furniture implies. In today's architecture, American architects have ideals about space, and in order to reveal more space, they show more floor by making furniture legs thinner. To arrange large spaces, architects suggest partitions to divide a room into two, and some partitions may be movable. Natural fiber and raw materials are also in high demand.

Most American houses are shrinking in plan. A large house requires more expense and has a lot of wasted space. The large lounge chairs of an earlier time are selling less, and getting smaller. Heavy dressers have been replaced by smaller and lighter ones, which are even lower in height. Given this trend, our furniture in America may lose its legs altogether in the near future. The interior of a room may resemble a Japanese futon, or something even more comfortable to sit on. The soft tatami floor may be regenerated as a more sanitary floor that is easier to clean.

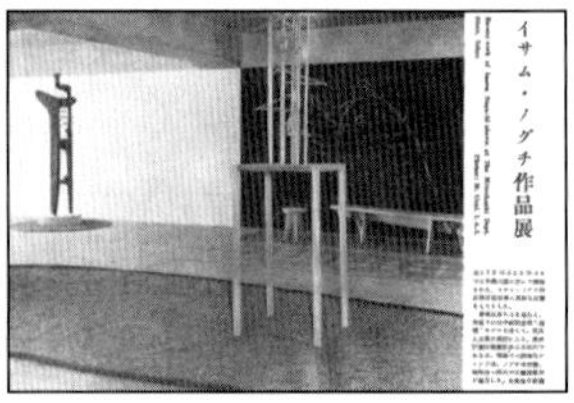

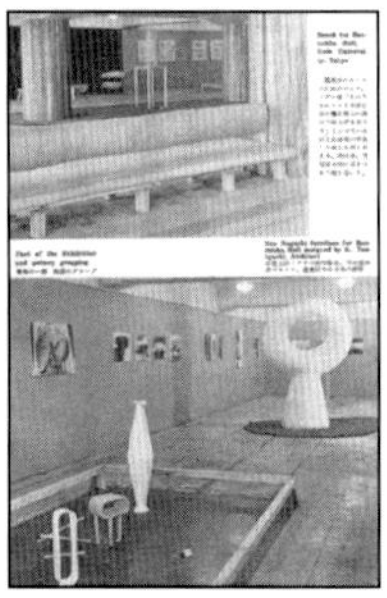

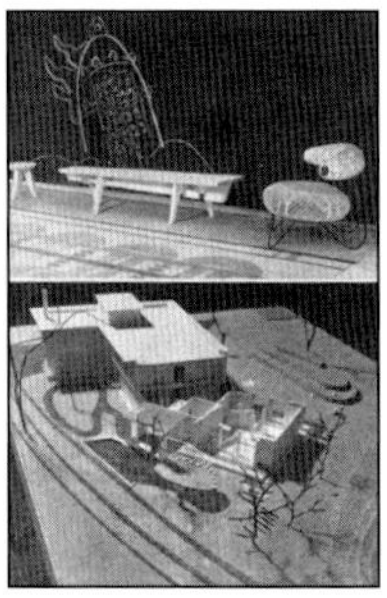

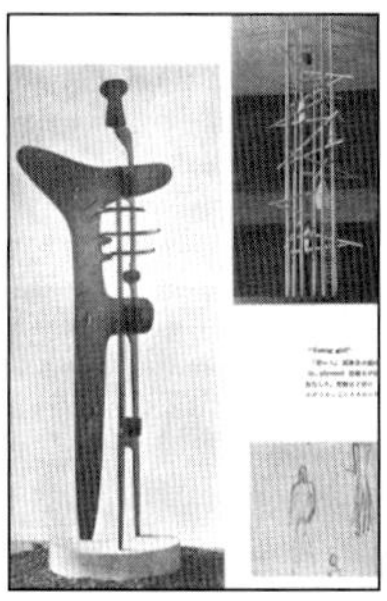

In the past decade the American popular lifestyle has changed greatly. There's no need to describe each and every transformation here, as they can be learned easily by flipping through magazines. The new trends and areas of research are easy to understand for the Japanese, and I think the Japanese have a lot to contribute to this trend. Furthermore, I think this trend has the potential to develop in a much purer way in Japan.

The Herman Miller Company

Most American furniture manufacturers are not fond of modern design. Herman Miller was one of them, when it was nothing more than a small ordinary company in Michigan. One day George Nelson arrived on the scene with a small new product he designed. This product turned the entire company around. This was not a bad thing for Herman Miller. They became aware that new designs can be beneficial to companies. This is perhaps one of the reasons for the birth of industrial design. As for those who make these industrial designs, they must be artists in a strict sense, while being great businessmen. Norman Bel Geddes and Raymond Loewy are both excellent businessmen and artists, and in fact, they became very rich, so rich that they could build their own factories. They hire young architects. They take away jobs from architects, leading to complaints that architects have suffered a decrease in jobs in the last decade. But they did not only design furniture; we all know Loewy's steam engines and Studebakers and Dreyfuss's streamlined locomotives. They may not work on their own, but they make us work. These people also know the psychology of the general public very well. I know Henry Dreyfuss, and he once asked me to collaborate with him. It was to create a new form for his Royal typewriter, and he asked me for some suggestions. After hearing my comments, he said he wanted to place a chrome tube around the machine. I could not agree with him, to which he replied that I did not understand the psychology of a typist, and we did not get any further with it.

Industrial designers are said to be the link between architects and the public. However, it is foolish to think that the public thinks less of industrial designers. To prove this, many nameless architects have switched to industrial design.

Design Education

Many of the new designs produced so far are grounded in demands. There are numerous industrial design schools throughout America that teach design fundamentals to supply the market. The Bauhaus in Dessau was based on a new method of education. After Hitler drove the Bauhaus away to the United States, both Gropius and Moholy-Nagy taught in America. Moholy-Nagy once taught talented youths at a school belonging to a cardboard container company. He passed away four or five years ago. When I visited him, I had the impression that he was creating a new trend in academism.

The key to learning the fundamental demands of design is to understand modern life.

ISAMU NOGUCHI AS I SEE HIM

From *Industrial Art News*, Vol. 18, No. 10, October 1950, pp. 19–23

ISAMU KENMOCHI

1.

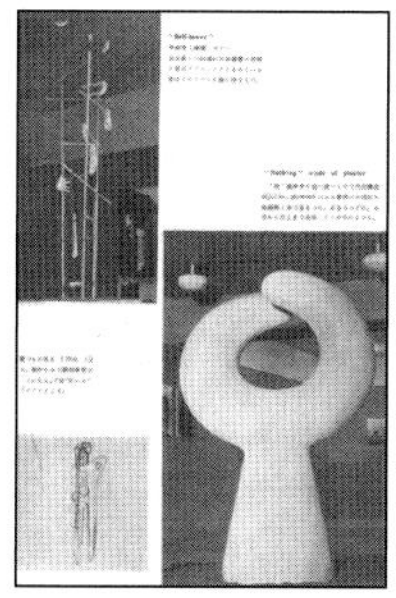

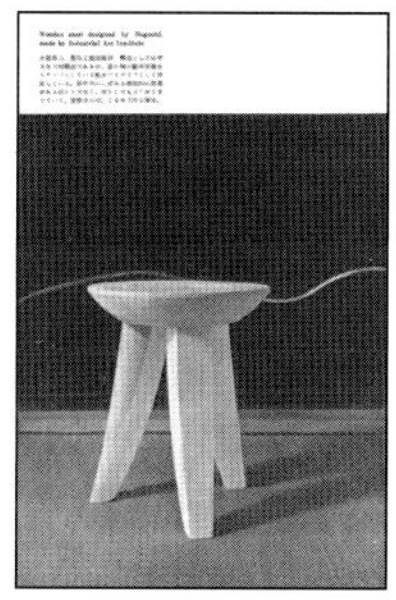

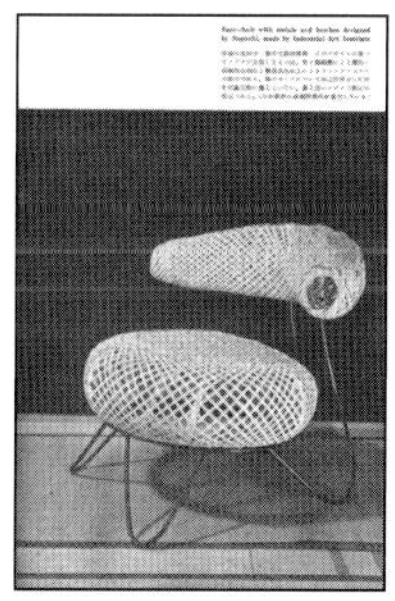

It was on a rainy day of June 24, 1950, that I met Isamu Noguchi for the first time. We were at Professor [Kenzo] Tange's office at Tokyo University's architecture school, and Noguchi had his thin figure wrapped in a gray raincoat. With his gray-eyed gaze, he said very quietly, "*Kon-nichiwa.*" It was hardly imaginable then that this man, once at work, would transform himself into something indescribable, perhaps a "work bug" or an "art demon," for lack of better words.

On July 28, Noguchi made his first visit to the Industrial Arts Research Institute, accompanied by his apprentice Michio, art patron Mrs. Greenberg, and Genichiro and Fumiko Inokuma. This time he was wearing a thick kimono without an undergarment, and *geta*—high wooden clogs. No other visitor, Japanese or not, had ever shown up in such attire, and all workers were transfixed by him, making their eyes round and small like a pigeon's. It turned out that his kimono had been ruined by rain the day before. He was unable to retrieve his own clothing, and thus ended up in such a style. On top of that he brought with him four chopsticks tied into an X-shape that he toyed with endlessly. As soon as he sat down in the director's office, he began abruptly, "No one has used this structure for a chair yet, right? Using this in a fundamental way, we can develop a design. Three-legged chairs are ordinary but . . ." Then he suddenly stopped, and the interview started, which was followed by a tour of the Institute.

In the modeling room, he made rather pungent critiques of our bamboo furniture and our plywood furniture, formed in a high-pressure mold. Some were positive, and some were in the form of a frown. In the gallery, he noticed a basecoat sample of lacquered metal, and said, "How beautiful. Why do the Japanese try to copy foreign countries when they have so many wonderful things? Especially when American architects and designers are trying so hard to grasp anything from Japan . . ." He was surprisingly impressed with the facilities in our wood and metal shops, especially upon being told that this was a government institution.

At the end Noguchi said, "This is such a nice place. Now I want to work here." And this is how from August 1, Noguchi began his daily commute to the Institute, wearing his blue round-neck t-shirt and blue pants, with a leather tool bag in one hand and a sketchbook in the other, with his apprentice Michio tagging alongside him.

2.

If you think Noguchi randomly uses assistants, you are mistaken. The man expects nothing from others. He once surprised me by saying very clearly, "I really don't need an assistant." Let me take you through his production process step by step, using the sculpture *Wakai Hito* (Young Person) as an example. First he drew lines at full scale on two pieces of graph paper, which were cut out with scissors and a knife blade. The two patterns were then copied onto two thick pieces of plywood.

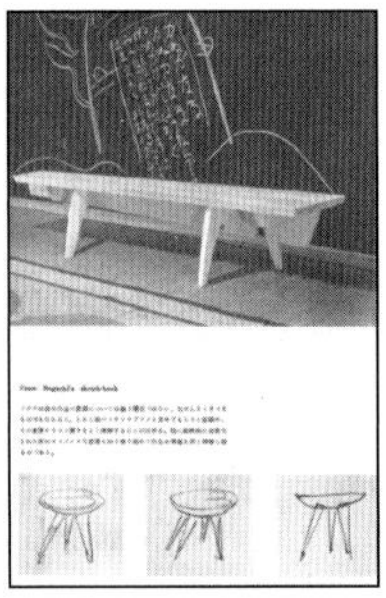

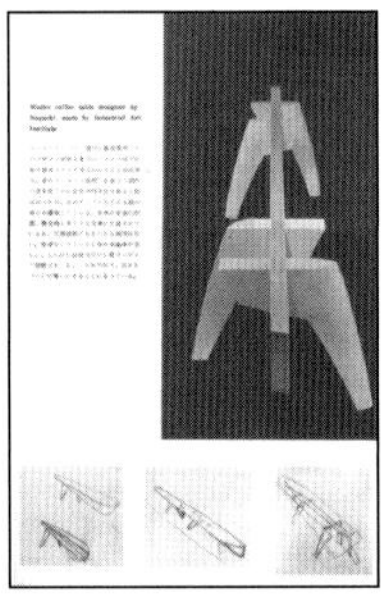

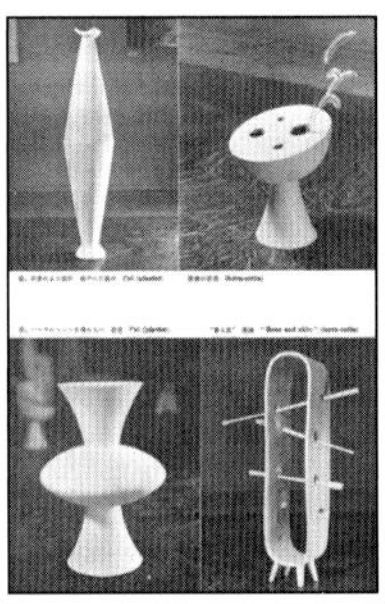

When our men offered to cut these shapes on the band saw, he clearly refused by saying, "No, I will do it myself." He checked the conditions of the Institute's band saw. "This is an old model," he said. This was the latest we had, though if compared to what was available in the U.S., I'm sure it was. As soon as he finished adjusting it, he swiftly cut out the curves. The edges were filed or chiseled to make sharp ridges with sculptural details. When the pieces were assembled, they fit together perfectly, and the beauty of the sculpture took my breath away. I asked an unnecessary question: "Is this human?" In response, he just gave a slow nod. It was then disassembled and given the lacquer basecoat finish, which he liked. (This was a half-scale model for the steel sculpture.)

3.

At this time he began numerous other projects, pursuing them all simultaneously as if he were a *Senju Kannon* (a thousand-armed goddess). By the time this sculpture was finished, he had already completed a wooden coffee table and a three-legged stool. These were both millworks, so we did the actual production, but Noguchi designed even the smallest detail by making full-scale drawings. His way of design and drafting seemed rather carefree: just like a Japanese carpenter, he used a very short pencil and a metal ruler, and quickly drew full-scale drawings for the coffee table. This is not to be mistaken with rashness, for one quick look inside his sketchbook will reveal how much thinking he did before he reached the final shape.

The coffee table top is supported by a long apron, with four alternating legs notched into it. The apron and each leg are connected using a lap joint, and because of the difference in thickness the notches on the apron are deep and visible. To this Noguchi said, "This cut-out is especially interesting." Each of the reverse L-shaped legs was joined in the double-tenon method specified by Noguchi, and was also visible. This confirmed my thoughts, collected in magazines and photographs, that in America, a simple and visible joint is one of the most distinct features in modern design.

The round stool was inspired by the ancient Chinese *shaku* chair. According to Noguchi, a shaku's form is not divisible by any number, and each leg has a unique profile. The legs for Noguchi's stool also had unique profiles. This seemingly small detail created a significant effect in the form. In particular, the curves and surfaces in the vertical and lateral sections were clearly not by a designer but by a sculptor, whose attention to precision and true form left no room for adjustment.

These pieces of furniture and his sculptures *Wakai Hito* and *Mu* (Nothingness) showed no discrepancy whatsoever between the lines he drew and the actual three-dimensional forms that were built. From similar experiences, I saw the accuracy of his instincts, and the careful detail of his three-dimensional planning. The depth, sharpness, and experience of his work were completely out of the ordinary. This cannot be attributed to talent alone, but also to the training of his senses. Once he starts work, all common social acknowledgments, meals, and sleep will be forgotten, and the only thing left is devotion to his art. This willpower, strength, and energy can only be compared to a kind of religious austerity. It made it all the more strange that he was an American. Sometimes I fell into an odd illusion that we were ordinary Americans looking at Noguchi, the Asian.

4.

Bruno Taut in 1934. Charlotte Perriand in 1940. And this year, in 1950, we welcomed Isamu Noguchi. First was the architect, followed by an interior designer. And Noguchi was a space artist. Taut already

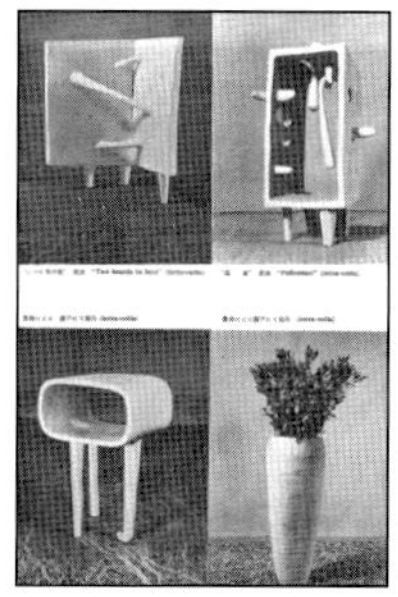

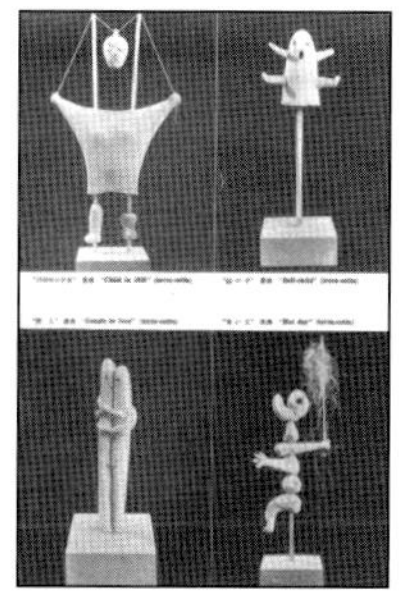

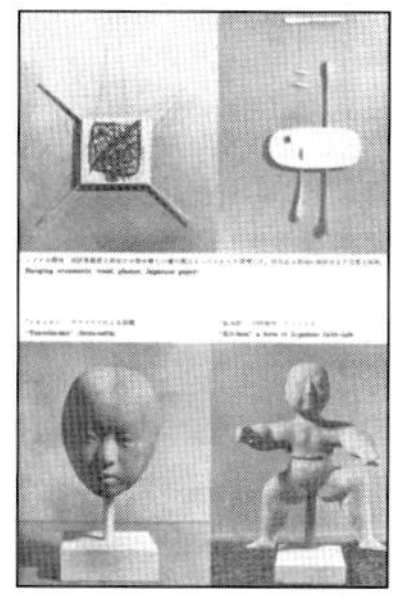

had a knowledge of Japanese architecture prior to his visit, and during his long stay he analyzed and critiqued Japanese forms, and suggested new possibilities based on Japan's fundamental forms, as well as its traditional materials and techniques. Similarly, Perriand was taken with the "functional tradition," something that can only be experienced in Japan, and showed us numerous pieces under her motto of "selection, tradition, and creation."

I found that Noguchi had a far greater ability to realize all of the above. I did not come to this conclusion just from observing his solo exhibition. Beginning with the architecture and industrial design magazines he was featured in, and learning about his career, then meeting with him for the first time, listening to his lectures, and hearing his few uttered words, I was convinced that Noguchi, out of all foreign visitors, is the one to truly understand Japanese art, who can make art founded upon Japanese soil and still be so international. And it's not only because his blood is half Japanese.

Those of us with long tenure at the Institute have been awarded the opportunity to see first hand the works of these three internationally renowned artists. At the same time we cannot help but compare the three.

Noguchi was a generation younger than the other two, making him closer to us in age, and the energetic sensibility that filled his pieces touched our souls, with or without explanation. It may be interpreted as a resonance of senses. In Taut's pieces we saw class and beauty, but we also could not help seeing them as a little outdated. The wood chairs he taught us to make were in a Scandinavian style with a hint of the nineteenth century, rather than the new laminated chairs. He did not even care for steel tubing furniture.

As for Noguchi, he also did not care much for steel tubing, but he recommended solid metal chairs, and was deeply interested in laminated furniture. "However," he would say, "flat laminated furniture is meaningless. They should be molded into three-dimensional shapes, so that form and balance can have a relationship."

All three, Taut, Perriand, and Noguchi alike, disliked steel tubing furniture for reasons of anti-humanism. They all liked wood or other organic materials, in opposition to shiny and cold metal. Taut was very technical in the details, rather than the overall form, and to us these seemed to be old tricks. Perriand and Noguchi were similar in that they created many solid configurations that were achieved by slicing solid woods. In fact, Noguchi saw the large table by Perriand in her catalogue, and said, "This is well done." But Noguchi's table was more relaxing to us, and especially to us Japanese. We felt as if it was our own. It had the same boldness and clarity found in the block style in Chinese calligraphy.

5.

Bamboo was another material in which Noguchi was acutely interested. When you follow the lineage of Taut, Perriand, and Noguchi, it is interesting to see that with each generation the use of bamboo as a material gained more and more understanding. Taut's use of bamboo bordered on that of rattan, which is a fixed-diameter material without joints. He drew sketches of bamboo furniture that to our frustration could not be made. By the time Perriand arrived, bamboo chairs were already available in our country, and she immediately set about using the material. Hers were modern forms, and she showed her original sensibility in the weaving of bamboo balanced with the folk use of round bamboo.

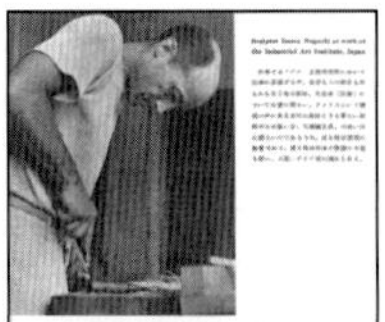

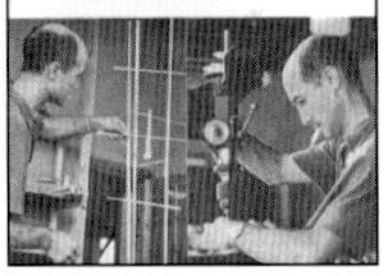

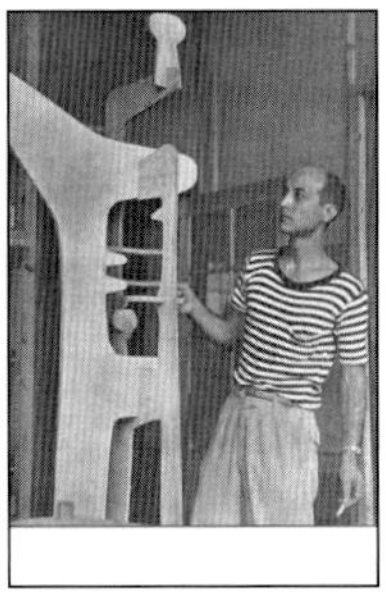

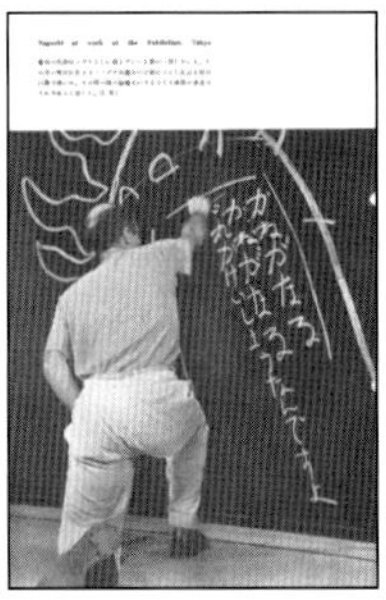

So far Noguchi has only explored bamboo as a material that can be woven. More than anyone else he understands the international value of bamboo, and he seems confident of its marketability in America. This is probably due to his having seen bamboo products after the war, and he has already witnessed its practicality and the value of things Asian in the marketplace. He says of bamboo, "There's no technique of weaving in America. This will be welcomed if we bring it from Japan. Since there is no indigenous bamboo to be found, woven bamboo will be met with enthusiasm." I've been saying that bamboo is making its way into American daily life, but according to Noguchi, bamboo from various Asian countries is already being sold there. Japanese bamboo is particularly known to be of the highest quality.

Noguchi's use of bamboo obviously concentrated on developing its pliable nature. Or rather, he did not care to use it any other way, beyond taking advantage of its elasticity. Perriand was the same, except that she laid pieces of split bamboo in a row. Noguchi did mostly weaving, as can be seen in his exhibits. The steps that preceded his bamboo chair are as follows.

First, Noguchi had the idea of making a stool by wrapping two intersecting pieces of plywood with rope. "The construction of a Japanese *tsuzumi* drum," he said, "can be immediately adapted for a chair. Let's make it with this idea." His mind, I thought, was packed with creative ambitions stimulated by seeing many different structures and forms since arriving in Japan. I asked for his critique of my chair prototype, a chair that was inspired by a Chinese wooden chair. He sat on it gladly, with an expression as though analyzing the long fat curve that extended from the back to the sides. Then I mentioned an idea I had had for a while: a chair whose seat is a basket. To this he said, "That is very interesting. Let's make this chair. Let's design it together." When he returned to the Institute the following morning, that unique form was already in his sketchbook. My original idea called for wooden legs, but his chair had legs made of solid metal bent to a gentle curve, a round basket seat, and a half-moon banana shape arm, which was a lot more continental than that of the Chinese chair. It was so creative that I tried not to compare it with my design.

These two ideas started from the same basic concept, but the difference in the designers' philosophy, sensibility, and environment can bear quite divergent results. That realization brought worries as I reflected upon my own work.

Back to the basket-seat chair. From the sketch, I made full-scale detail drawings, and a model of wire and clay was made for further analysis. I bent a curve for the leg that was quickly bent further by Noguchi's large, forceful hands. Instantly the curve acquired beautiful line and function. For the round seat and the banana-shaped back, Noguchi suggested, "Let's make them thicker. That would make it look more comfortable, and it actually will be." So another layer of clay was added. From the drawing and this model, a solid iron leg was produced immediately. As we only had half a day for production—and the sun was blazing—the workers in the Institute's metal shop were drenched in sweat. The bamboo parts were delegated to Harumitsu Inoue, a bamboo engineer. Inoue tackled this entirely new experience with neither sleep nor rest. I asked him to make and remake the shapes until they were satisfactory. I wanted to convey Noguchi's "no compromise" principle.

The morning the chair was completed, Noguchi showed up on the hill of the Institute, not in his regular t-shirt and wrinkled pants, but all dressed up.

He immediately sat down in the chair, and moved his upper body front to back, left to right.

"Not bad!"

Of course it was not perfect. Even we could point out a few imperfections, whether of design or craft. "Of course," he said. "Furniture, and especially chairs, cannot be a success the first time. Everyone learns from the first model!" Having decided to add this piece to his exhibition, Noguchi happily promised perfection on the second model. "The two holes on the arm's ends are really interesting." These holes were not anticipated at the time of design, but were an accidental yet fortunate outcome. According to Inoue, when you start weaving from the center, these holes cannot be avoided at the end.

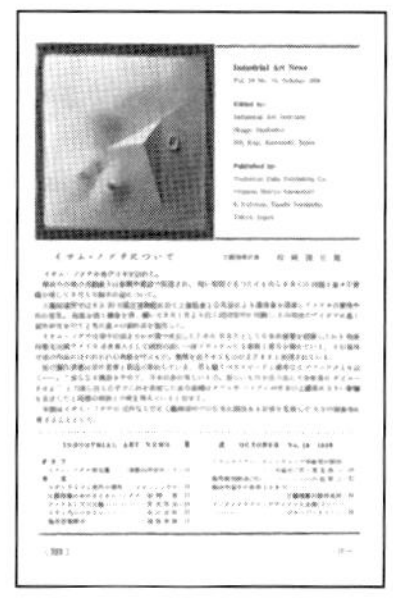

The chair was brought outside, the morning mist still in the air. Noguchi brought out his camera, and took photographs from every possible angle. Many shots were captured with his favorite Leica, the one we knew that even Inokuma was not allowed to touch for five years. "I'm going to show this chair to George Nelson," he said. "I think this chair will sell in America. I'll talk to Herman Miller, please make lots and send them." We attempted to make two or three more bamboo chairs, without success. However, we promised to complete the basket-seat bamboo chair.

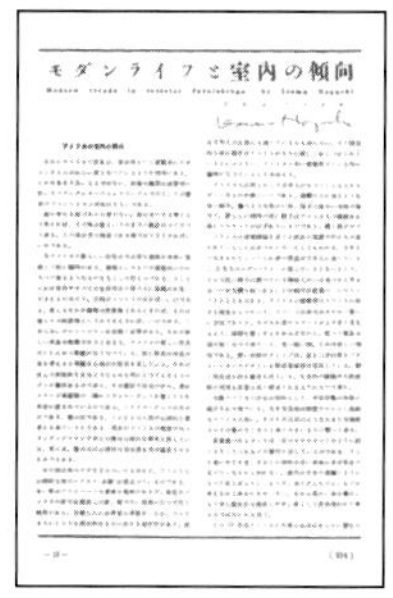

モダンライフと室内の傾向

6.

Noguchi's days at the Institute were intense and productive, and lasted two weeks. Two seven-foot tall sculptures, three pieces of furniture, two models of a tower, *haniwa* pottery, a vase, a hanging scroll, and probably close to twenty additional pieces were made during this time. What did we learn from him?

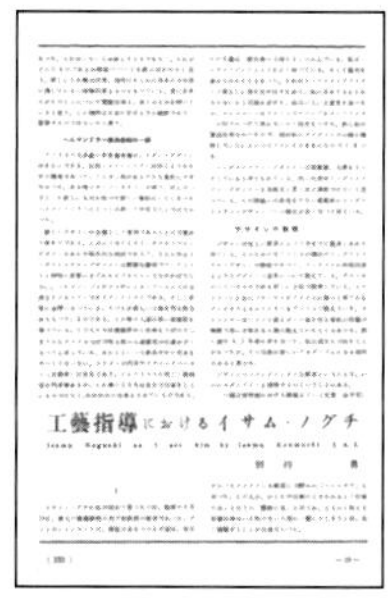

工藝指導におけるイサム・ノグチ

The biggest lesson was . . . that we Japanese were Tyltyls and Mytyls, chasing after the Blue Bird.

Noguchi loves primitive art. Pointing to ancient Chinese characters and block calligraphy, he said, "Abstract art has always been around." He praised the Balinese god carvings and handcrafts, and enjoyed the Japanese *haniwa*, old *koksehi* dolls, and paper maché masks. He admired the structures of the past. Not only the Ise Shrine, Katsura, or Ryoanji, but his photographic collection tells us what he was attracted to during his travels through the ruins in China, India, Indonesia, Greece, Spain, and Italy. What he observed and was moved by will eventually be released as his own art. He will ignore all preconceived rules, concepts, or ideas, and will simply create, with strength and passion.

Could we do the same? Not so easily. There is a major difference between Noguchi and us. He has a broader and higher outlook that is worldly. The more you remove yourself from a thing, the more accurately you will know its value. Noguchi sees the world's cultural heritage from a high viewpoint, and calmly makes a selection. He will not misread or overlook. The past gets renewed through his originality as totally new art.

We should learn to see the "references" we have, such as tradition, climate, or our own materials, from high above and think in depth. Twentieth century journalism has shown us—even those of us stuck in Japan—endless streams of information to prove this. It is important to think broadly before we begin our design process. I remember Taut's words: "The architects in Ginza should learn to think like the Bodhidharma."

7.

A week before his departure Noguchi told us, "The East should learn the West's civilization. The West should learn the East's cultures. And only in a world of freedom can our countries naturally develop relationships."

THE WORK OF SCULPTOR ISAMU NOGUCHI

From *Industrial Art News*, Vol. 18, No. 10, October 1950, pp. 24–25

MOSUKE YOSHITAKE

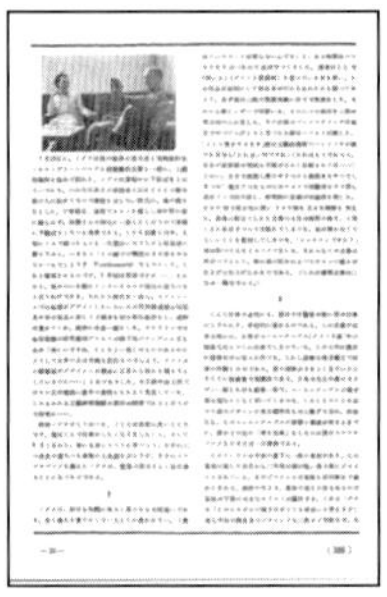

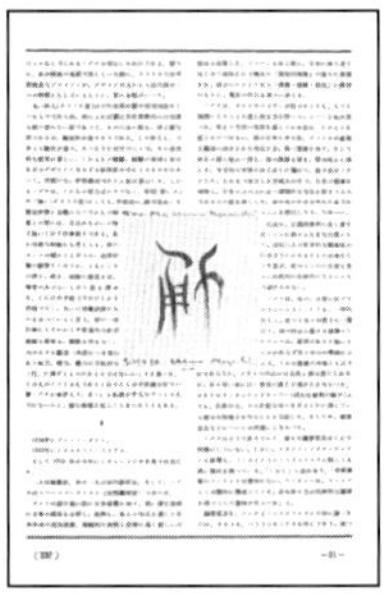

Toward the end of Isamu Noguchi's stay in Japan, I had the opportunity to witness first hand this renowned artist's production process. Here I'd like to share a few thoughts I gained from the experience.

Of all fields in the arts, I think the world of industrial design was most affected by Noguchi's visit to Japan. I believe this is because his abstract sculptures are designed and shaped with relentless planning and scrutiny with regard to materials, fabrication, and finishes. How the word "design" can be applied to art or sculpture is clear once you see the refreshing works by Noguchi. For designs the entire industrial world relies on, preoccupation with superficial aesthetics would be unbalanced and unacceptable. Here the immediate problem is compromise with commercialism, with aspects of mass production and marketability. But if it is a creation of form, lack of beauty is also unacceptable. Industrial design consists of various compromises, with esthetics as its foundation. In this regard, the industrial designer becomes a production staff member, and will be responsible for its profitability.

Noguchi's specialty is in art, not industrial design. However, his meticulous care for the elements of the sculpture being assembled—for its composition and the finish selected specifically for the material—and his consideration for the final installation environment all emphasize an adaptable and perceptive nature that is far from yesterday's sculptural philosophy, which advocated a microcosm of pure art. I believe this is what nurtures the aspects of Modernism in his work.

Noguchi's adaptation to the functional aspect of objects has inevitably taken an abstract form. It was no surprise that his abstractionism created his furniture, stage costumes, and set designs. Through multiple journeys, his art pilgrimage brought a reaction to the style defined by Rodin, and through an admiration for primitive arts he arrived at abstraction. The sculptures he designs have a new idiom of forms realized through a blend of adaptation to the environment and new industrial technologies.

Having seen and heard Noguchi, it is quite interesting to compare him with the two other foreign designers the Japanese industry has welcomed in the recent years: Bruno Taut and Charlotte Perriand. Upon finding the same rationalist principle employed in Bauhaus architecture in the Katsura Imperial Villa and the Ise Temple, Taut was shocked and elated, and called it "the precept." He calculated the master model dimensions for chairs, tables, and lamps using the standard body measurements, and set us at the starting point of an industry that had had no standard whatsoever. It was a numerical treatment for the foundation of design.

Perriand was invited during the early half of World War II, when the export industry was prospering. She took up the task of making new achievements in design. Her instructional theme was "tradition, selection, creation," and her selective eye would choose a particular industrial technique that was then made into various designs with a new sensibility. We were amazed and embarrassed by the skillfulness of these pieces created from our own land and climate. She showed how design can be practiced using the new industrial philosophy that followed Taut's teachings.

Noguchi focuses on beautiful creations as a sculptor who designs. But is he a sculptor only?

Caught up in endless debate over "standards" and the repetitive analysis that results from rationalism, our industry seemed to have declined in recent years. But Noguchi reawakened us to the pursuit of true beauty. And the beauty he showed us is bold, refreshing, and liberated; it is full of Modernism. We all realized that we were losing ourselves in half-baked logic. It was time to realize that what we create must be beautiful.

Noguchi's furniture design is grounded in material composition. He does not dwell on economic considerations; his focus is on beauty. Each of the four legs on his coffee table is similar but different from the other three. I gathered that this came from a firm visual grasp of an animal's hind leg, and it made the table so much more beautiful. The legs that supported the woven bamboo seat [designed with Kenmochi] were made of thin steel in a long and luscious curve. He made an effort to create them so.

From its skeleton form to plaster finish, his ten-foot tall sculpture *Mu* (Nothingness) was completed in just one day. The speed itself was unheard of, but more astonishing was the way Noguchi, half-naked and splattered with plaster and sweat, would one moment jump onto the scaffold with plastered hay, and jump off the next. It was a fiendish spectacle in the dead of night.

As soon as he was done with the plaster, he was already on to the next design. His desk was covered with ruler, compass, scale, slide rule, and patterns. It reminded us that art does not come from leisure.

Mr. Nan-e from NHK precisely described the relationship of Noguchi's work to an American context in his introduction to Noguchi's lecture at Mainichi Hall. According to him, Jean Cocteau wrote in his travel notes that America has comfort but no luxury. However, Noguchi would skip comfort and aim straight for luxury. American comfort is the public standard, and European luxury is its apex reached through investigation. This clearly explained Noguchi's nature and discipline.

The significance of Noguchi's visit to Japan, to me, is as follows. At a time when the Japanese industry, still in the thrall of Taut's rationalism, was beginning to veer off track and lose its aesthetic bearings, Noguchi as the sculptor who designs gave us a warning: Industrial arts must be beautiful and new.

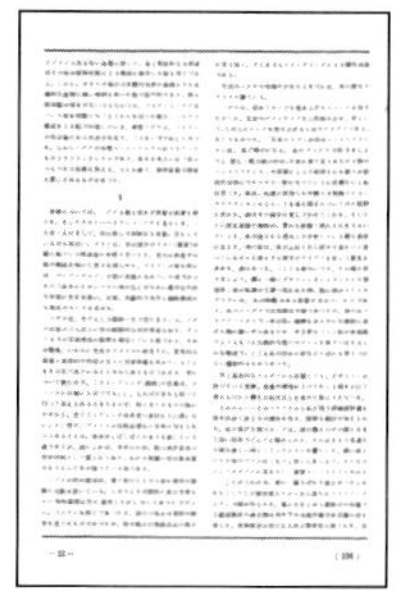

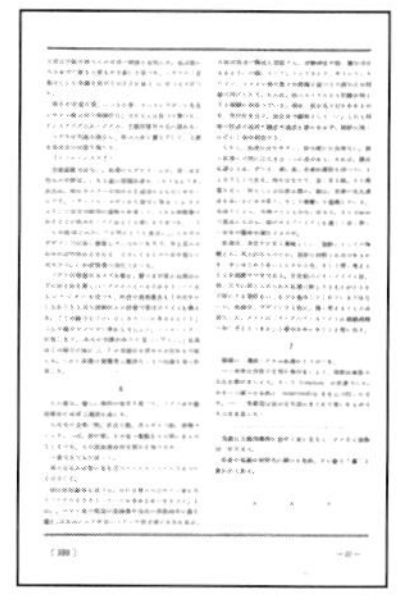

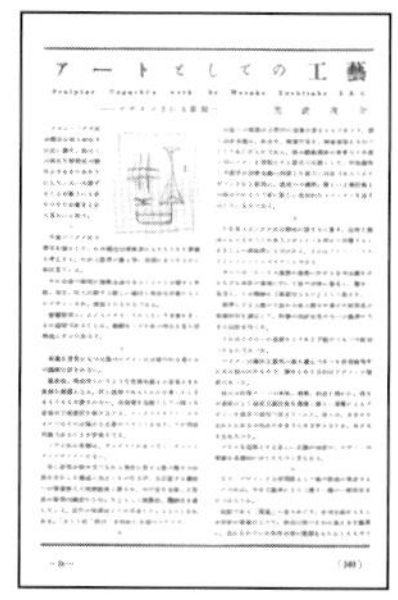

ISAMU NOGUCHI'S DESIGN

From *Industrial Art News*, Vol. 18, No. 10, October 1950, pp. 25–26

YOSHIRO TANIGUCHI

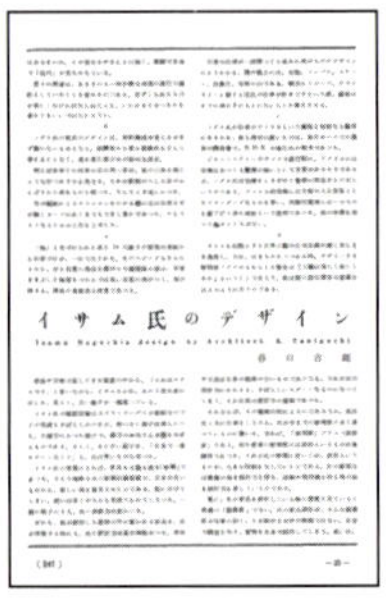

イサム氏のデザイン

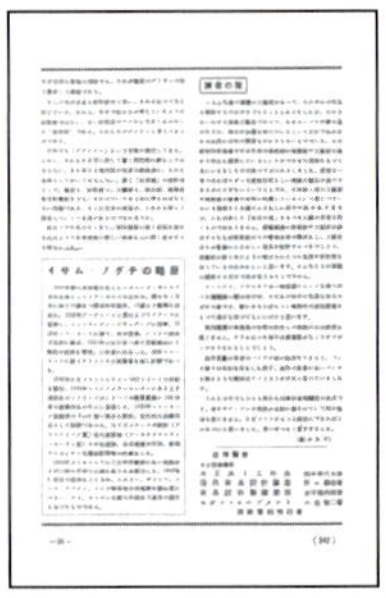

From among the photographs he took in Nara and Kyoto, Isamu chose one and handed it to me. "This is nice," he said. It was a photograph of an old chair.

Isamu's photography always impressed me with its fresh angles, but this chair was especially beautiful. It was a chair he had found at Daitoku Temple in Kyoto, for Zen monks to sit on. Smiling, Isamu called it "the most modern chair in Japan."

According to him, all furniture, crafts, and gardens are sculpture. His unique sculptural eye rediscovered the Japanese beauty within the old. To an ordinary eye, they were already old and forgotten, yet somehow his sharp insight could reach deep into the past to detect this chair.

While he was at work on furniture commissioned for the building I was designing, he showed a design ability that was quite unique. However conventional the materials and techniques may have been, once touched by Isamu, they all turned into beautiful modern designs. This was the magic of his design.

So what made this magic? During our collaboration I was astonished to see how different he was from all of the other sculptors I had known. He was, so to speak, a combination of sculptor and designer. Until now a sculptor always detached himself from design, but in Noguchi's sculptures, design plays a large role. His ability to sculpt is at the same time an ability to design, in the way that design is practiced in the making of architecture, ships, and airplanes.

Furthermore, observing Isamu's attitude while designing furniture revealed that he is no ordinary "sketch designer." By this I mean that his furniture designs do not stay on paper only. He creates models and produces prototypes on his own, or he may take charge of the production team on site, and this process is as fast and accurate as a *dessin* made by a painter.

Noguchi will always call himself a sculptor—a special responsibility he has taken upon himself—which makes him a very different sculptor from the ones we have known. He has moved far beyond the sculptor's realm, to be an artist of forms. This I shall call a designer.

Here in Japan "designer" has recently become a very popular word. However, we cannot just leave it as a mere trend. Nor should we attribute the word solely to those who make cheap copies of fashionable clothing. It is an issue for all painters, sculptors, craftsmen, producers, architects, and city planners to rethink their approach, as those associated with a lifestyle of beauty. Here lies a path for the further development of Japanese arts in the future.

I was greatly stimulated by Noguchi's vital attitude toward production, through which I deeply felt the new mission bestowed upon the arts in Japan.

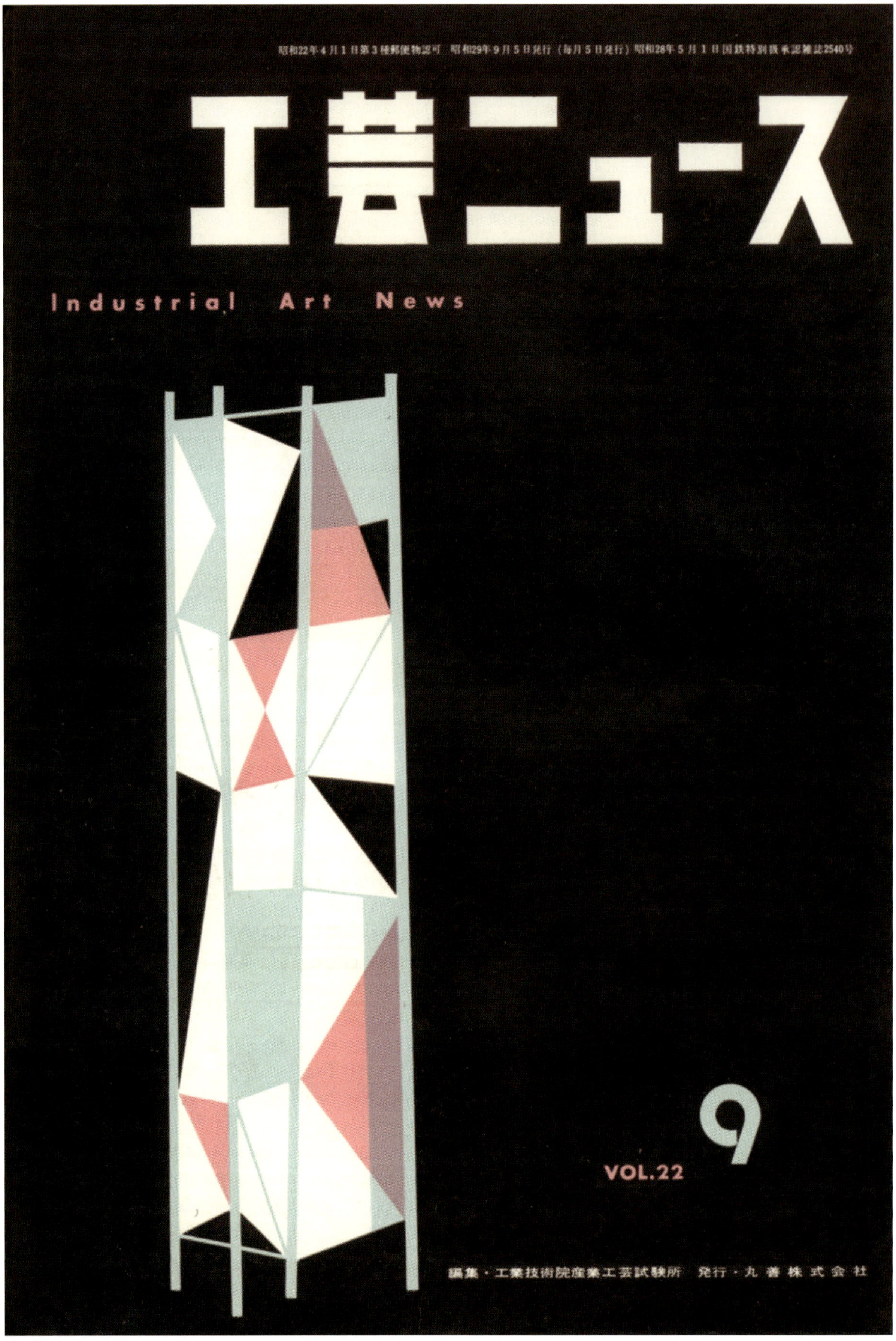
昭和22年4月1日第3種郵便物認可　昭和29年9月5日発行（毎月5日発行）昭和28年5月1日国鉄特別扱承認雑誌2540号
工芸ニュース
Industrial Art News
VOL.22
9
編集・工業技術院産業工芸試験所　発行・丸善株式会社

JAPANESE MODERN AND JAPONICA STYLE:

TWO ROADS TO PUT JAPANESE INDUSTRIAL ARTS ON FOREIGN EXPORT

From *Industrial Art News*, Vol. 22, No. 9, September 1954, pp. 2–7

ISAMU KENMOCHI

The controversy over "Japanese Modern" has been even more heated recently. Evidently I am responsible for initiating the use of this term. Since the end of 1952, I have argued that exports from Japan should not be just a B-grade humiliation export, but should be high-quality exports of Good Design. With today's export tactics, we are bowing our heads lower than we need to, and degrading ourselves too much. Even the large trading companies are satisfied with exporting second- and third-rate goods. However, American citizens treat our products with much more respect. They are asking for "outstanding Japanese products" instead of cheap third-rate products for the general public. Just as high-quality designs from Sweden are called Swedish Modern Design, ours should be referred to as Japanese Modern Design. I know that a top trading company in New York is "wide open" to Japanese modern designs. Design from Japan—that is, outstanding and high-quality design—is what I meant when I started using these words.

However, quite unfortunately, the term was taken in a superficial way. I feel as if I was trumped by those who claimed "what a shallow idea it is." But I did not give enough explanation to avoid being trumped either. First, the Japanese translation for "Japanese Modern" as "Modern Japan Style" was not good at all. These words remind us of others such as "new Japan music," or "*Taisho-koto*" (young readers may not know the latter, but this was the cheap version of the Koto instrument popular in the Taisho Era), or a poorly designed stage set for *Madame Butterfly* . . . or anything as shallow as these. In other words, it seems that the words Japanese Modern were conveyed and understood as an exotic fashion in America, the kind represented by Fujiyama and Geisha styles. I would like to point out here that for the Fujiyamas and Geishas, we already have another word, Japonica. And we should not be confusing Japanese Modern with Japonica. Japonica is a rather derogatory word, while Japanese Modern is a normal and serious phrase, meaning the "design of contemporary Japan."

Anything that refers directly to Japanese tradition deserves the title of Japonica. For example, if you had samurai-style hair (*chonmage*) and walked down Fifth Avenue in New York, you would certainly attract a lot of curious eyes. This may be the ultimate form of Japonica. The Japonica trend does exist in America. Some unknown designers may actually follow it. My use of Japanese Modern never pointed to such things. There are stores that specialize in Japonica in San Francisco, Los Angeles, Chicago, and New York. Any store that sells *Japan* can be categorized as one, although the ones that sell Japanese culture are different. The souvenir stores found in Yokohama, Kobe, or any military base sell Japonica. There, as you know, you will not find anything that was designed. There may be a few items that were contrived by merchants, but nothing that was designed. Instead you will find piles of cheap, shiny fake antiques, flimsy lacquered music boxes, and metal lacquerware covered with decoration. Actually, these Japonica objects are easy to spot, even to a novice. But more substantial Japonica becomes a little tricky to point out, and requires a lot more caution. The recent surge of industrial art exhibitions and export craft design competitions shows a great deal of this kind of Japonica. They all use techniques and styles that can only be found in Japan, which is their main asset, and even with inferior design, people fall for them. The design itself may be a disaster, but it may make excessive use of bamboo or employ an Asian technique that involves minute detail, and those making judgments eventually take the bait. I imagine that those who make them, in bamboo,

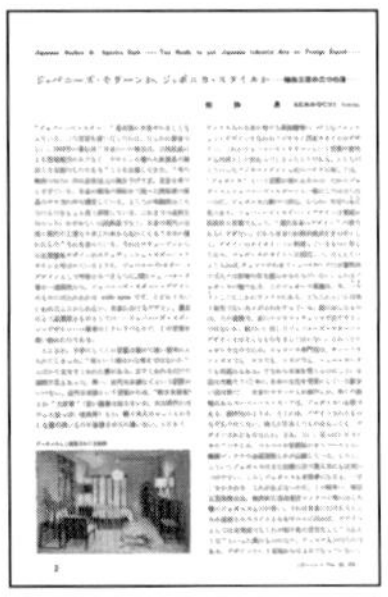

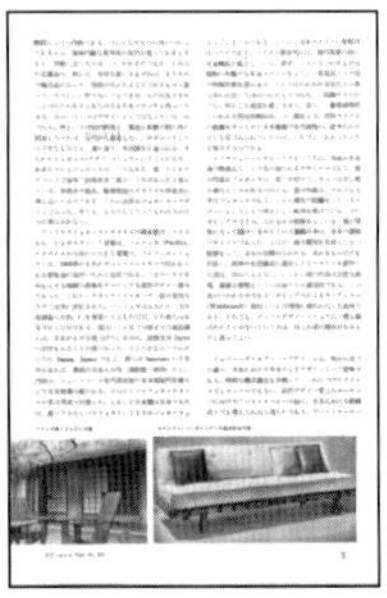

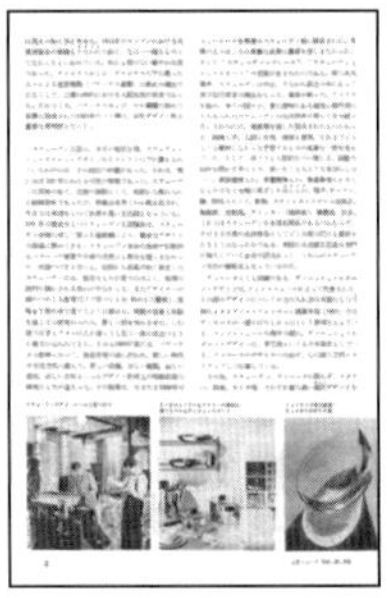

for example, believe that these qualify as great exports and will sell like hotcakes. They are eager to show that there is nothing they cannot make out of bamboo, so they make all sorts of things, but there is no design to speak of. None of them has considered the material characteristics of bamboo, there is no relationship between structure and form, nor is there any kind of modern feeling. They are not good designs, but they are selected for international exhibitions, which should be called Japonica exhibitions.

There is plenty of Japonica to be found in furniture as well. I remember a while ago there was a "National Style" furniture show held every year in one of the department stores. This was basically Western furniture crammed with the trappings of Japanese furniture, fittings, and the *sukiya* style. It was certainly Japonica. Even today, it's not hard to find.

In America, Japonica style has its roots in Hawaii. The word Japonica is derived from the Pacifica style, named for the Pacifica Show held in San Francisco in 1952. For the most part this was an exhibition of designs that had local scenery as themes, mainly Hawaiian. The show was introduced nationally with the help of *House and Garden* magazine. Pacifica obviously hails from the Pacific islands, but most of what was shown came from Japan. It's interesting that until recently the magazines that featured Japanese designs always described them as Pacifica, and we hardly ever saw the word Japan. But these days all we see is Japan after Japan. For example, the July issue of *Interiors* seems to be a special Japan issue, as the cover photo shows a piece by Genichiro Inokuma, and inside there is an article about the Japanese pavilion at The Museum of Modern Art. At any rate, the name Pacifica has become prevalent, but if it has anything to do with Japan, it becomes Japonica. The entry point for Japanese styles is Hawaii. Japanese style is infesting most new houses being built in Hawaii, from shoji screens and *fusuma* sliding doors to the exteriors, which could could be mistaken for an inn at a hot springs resort. For us Japanese, these cannot be endorsed. They are imitations without nationality. From the Rokumeikan Era at the beginning of the Meiji Period until now, we have been incorporating Western styles into our Japanese architecture and interiors. Now we finally know how Westerners felt when they first saw our versions.

T. H. Robsjohn-Gibbings is a designer known for his Japanese-influenced works. I regret placing his work in the category of Japonica, but it is nevertheless Japonica well done. He has been criticized for being modern through copying styles, yet he has studied extensively. He showed me his library, which was full of Japanese books. His purpose was not to duplicate, but to extract only the necessary out of Japanese traditions, and he created a style that fits the Western lifestyle. This deserves to be commended. However, the expression of this work is not always straightforward and honest. This could be why his company—Widdicomb, a very well-known company—has never been selected for the Good Design Show.

Japanese Modern design, as I said at the beginning, means good contemporary design in Japan. It is not a title or a slogan intended to establish a new style. It should not be mistaken as a new style from Japan, as Art Nouveau was a new style from Europe. Art Nouveau came and went like sparks from fireworks. All of the spectacles at the London World Exposition in 1851 left nothing in the end. They were all elaborate flowers without roots. On the other hand, the Bauhaus movement by Gropius and his followers was not an establishment of a style, but the pursuit of a new lifestyle in the industrial era. This is why the Bauhaus and the 1925 Paris Exposition that introduced Bauhaus to the world are both important benchmarks in modern design history.

Swedish industrial arts came a long way to achieve the status they enjoy today as Swedish Modern Design, and this was not without profound effort. It is the fruit of a century of research. It is well known that Sweden had long been feared by other countries as the Great Power of Scandinavia, but following its defeat in war, it began the process of democratization early, and it has now become a welfare state with a great culture. The Swedish Society for Industrial Design has been around for more than a hundred years, and has a nationwide network that strives to impart wholesome designs. It also collected and analyzed conventional technologies and traditional patterns. Although physically part of the continent, Sweden had the isolation of an island, which helped it to develop original designs without being affected by trends from other countries. Its designers were trained in the field, not educated in theories alone, and were encouraged to pursue new forms only through practical techniques and experiences.

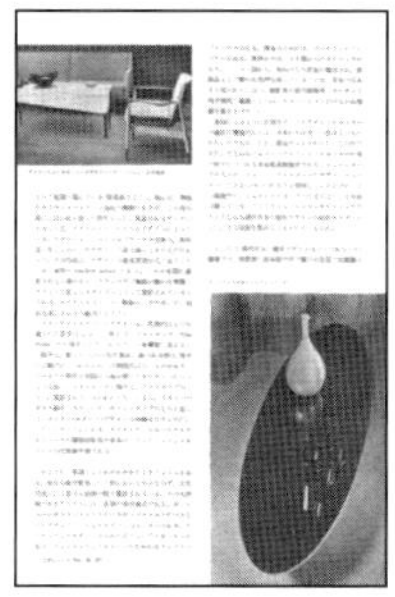

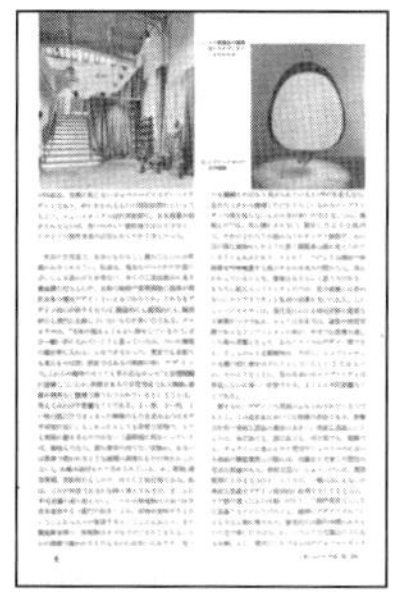

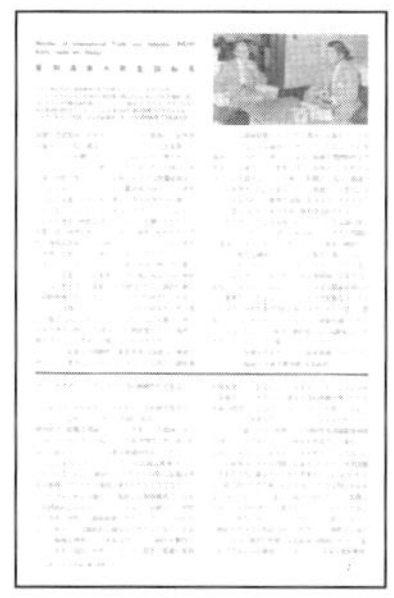

In this country, it seems that everyone thinks and works similarly. Since the 1930s, under the influence of the Bauhaus, functional studies were actively pursued, prompting research into the materialistic aspects of design planning, such as new product types, new functionalities, new technologies, and new materials. The results were introduced at the Swedish Pavilion at the 1939 World Expo in New York, and the world did not hesitate to shower praise. This generated words such as "Sweden Grace" and "Swedish Modern." During World War II Sweden kept its neutrality, in other words, its isolation, which provided another opportunity for self-reflection. And when the war was over, its newly improved housewares astonished many countries, including the U.S. They have now transitioned into perfection. Machine and hand, human and natural, intellect and emotion: these opposing values were unified beautifully. Above all, each object has the humanistic elegance that would make anyone, no matter what nationality, want to pick them up and start using them. Neither exotic, antique, nor mere novelties, they are real things based in a real land. Chairs, tables, shelves, light stands, rugs, stainless-steel utensils, porcelain, leather, wicker, textiles, and toys—all came to be loved in all countries. In the U.S. there are many trading companies that specialize in Scandinavian design products, feeding Swedish culture into the country.

The same can be said for Danish Modern Design. It is represented by Finn Juhl, and has received great national support. It is said that at the Ninth Italian Triennale (1951), Denmark won the best reviews. As we can see in the chairs by Finn Juhl, Danish modern design has its foundation in the country's handicrafts. Designers in Denmark always collaborate with the craftsmen they trust.

Other countries such as Italy, West Germany, and Switzerland export high-quality modern designs to other countries. One example is a thin, white dinner set from Royal Berlin Porcelain Company designed by Wilhelm Wagenfeld. Having spent nearly thirty solid years in design, beginning at the Bauhaus, Wagenfeld's works are strict applications of design principles. It is truly "qualität arbeit" (quality work), and easily accepted by other countries. If the Swedish porcelains are loved for their modernism, rich sensibility, and witty aesthetics, these Royal Berlin porcelains are attractive for their cool and intellectual beauty.

As for Italian Modern Design, it is not difficult to find its best examples. One is a chair by Gio Ponti. It looks very delicate at first, but is incomparably sturdy and lightweight. This elegant form is far ahead of us, who have just begun our own chair designs. The chairs of Charles and Ray Eames are found in Europe, but I do not know how popular Gio Ponti's chairs are in the U.S. Also at the core of Italian Modern Design is its glassware, which, like Swedish and Finnish glassware, is of the highest quality. And from Finland, we can name Tapio Wirkkala and Ilmari Tapiovaara as the foremost designers for their layered sculptures and furniture.

As you can see, each country has its own great contemporary forms beloved by other countries. This is not a form of export but of cultural exchange. America provides a major market for these countries through numerous import channels. Bonniers, Frank Brothers, and Hanna Elizabeth are centers for Scandinavian modern design. Georg Jensen imports Danish Modern, Finland House specializes in Finnish modern, and Gordon Fraser represents West German designs. From Asia, Thailand exports to America through Thaiback. It is little known in Japan that Thailand's refined products are exported to the U.S. and winning praise as high-end. For interior textiles—draperies and upholstery—Italy and Thailand are among the best.

Although many countries have their own Good Design centers in the U.S., we have not had a Japanese center until recently. It is called the Japanese Trade Center, and since I have heard that New Yorkers were anticipating the arrival of a center for Japanese design, I can only hope that this will not become a Japonica Center, and that it will become a place to introduce Japanese Modern Design.

Let's take a look at Good Design in Japan right now. In architecture, the Miyagi Residence by Kiyoshi Seike, Kenzo Tange's own residence, and Kenzo Hirose's work are all examples of Japanese Modern Design that we should be proud of. They are also womb-like spaces for the Japanese Modern pieces they will contain. If The Museum of Modern Art were to exhibit Japanese architecture, it is my wish that they show more than the traditional Shoin style architecture, and include these contemporary houses.

In furniture and houseware, where can we find the great Japanese designs? Sadly, we must admit that they are few and far apart. Should the things we find in most craft shows—jewelry, painted porcelains, or luxurious Makie lacquerwares—be called good contemporary design? When we analyze their design, most do not have a contemporary basis in function, feeling, or economy. As he was leaving Japan, Gropius is said to have commented, "Japanese bowls have such great shapes, I wanted to buy a set to take with me. But I could not find any set without a pattern. In Tokyo or in Kyoto, everything I saw was crafted with great skill, but they were all over-decorated, and I did not dare to buy them for myself."

I agree that the porcelain and lacquerware that we use daily are decorated without exception. This is indeed very strange. It is probably next to impossible to find solid-color dishes with good shape, color, and texture. And if we did find them, they would probably be abhorrent junk. A plain lacquerware bowl, for example, is likely to be either low-quality rubbish or something astronomically expensive for use in a formal tea ceremony. Most others come with pictures and patterns. The same can be said for rice bowls, tea cups, and tea sets. They all have pictures on them. I often wonder why, and I don't have an answer. One reason may be that in Japan, a plain item may resemble something ominous, or something cheap. It may be that there is an economic reason, given that manufacturers can hide imperfections behind the decoration. Some of us, who have only been exposed to these decorated dishes, may not even know the beauty of simplicity. Or could it be that in the buyer's way of thinking, plain ones seem to lack something—you aren't getting your money's worth, and you feel better when you buy something with more decoration? Whatever the reason, I find it worrisome that the Japanese seem about to lose their appreciation for simplicity. Of course, I do not think that simplicity alone is the foundation of beauty. There are opposite qualities that can be just as effective. However, we must remember that simplicity is a major element in contemporary aesthetics. It was once the most common notion in Japan, and now it is accepted throughout the world. Even in France, Louis XVI styles are being overtaken by simplicity. And yet, there is no such thing as simplicity in Japan. It is indeed very puzzling to me.

In short, I believe that the principles of Good Design are being destroyed. This is largely due to the influence of our arts and crafts. Whether metalwork, lacquerware, bamboo, or porcelain, the relationship between the pieces displayed in a gallery and the pieces sold as merchandise is similar to that of a member of parliament and his people. Every movement in the upper echelon of arts and crafts is reflected in the manufacturing world. However, at the risk of generalizing, there isn't much that makes sense from the standpoint of design. I have heard terms such as "industrial arts as pure art," but I have never seen a piece that was designed purely. In terms of aesthetics, the industry is limping. So what should people do with MP's like this? It becomes apparent that it is almost impossible to wish for Japanese Modern Design.

Does Japanese Modern Design deny tradition? The answer is no. It takes tradition into great consideration. However, it does not care to follow traditional styles. Rather, it looks to the principles that have been inherited in Japan. These principles will take different shapes, just as technology and materials change through time. Tradition as a principle is very important, but if we involve ourselves only in the stylistic aspects of tradition, it will hinder our future development.

Isamu Noguchi's lamps were inspired by Japanese paper lanterns, but there is no copying of the traditional style. Only the material, techniques, and function have been extracted, put into an original contemporary design. This is born of the Japanese climate, and it is an industrial design that can only be made in Japan. It is only natural that these lamps are quickly gaining popularity in the high-end market in Europe and America. And the prices are not high. You could probably purchase two or three of these lamps for the price of one vase on a department store's arts and crafts floor. No wonder it was selected for the Good Design Show, held this May in the U.S. The furniture by George Nakashima, who received an AIA award in 1953, is a great example of Japanese Modern Design. Both Nakashima and Noguchi are Japanese with American nationality, but what about Japanese designers with Japanese nationality? I could name a few here, but I will not mention them on paper. Instead, I propose a Japanese Good Design exhibition, for all Japanese and foreigners to see, and the exhibition should travel abroad. As long as the pieces are selected carefully, I believe such an event would bring great success to both export promotion and the presentation of our culture.

Texts above from *Industrial Art News* vol. 18, no. 10, October 1950, and vol. 22, no. 9, September 1954, translated from the Japanese by Eiko Sakai.

WORLD EVENTS AND ISAMU NOGUCHI–ISAMU KENMOCHI TIMELINE

This chronology was compiled from various sources, including Hitoshi Mori ed., *Japanese Modern: Retrospective Kenmochi Isamu* **(Matsudo, Chiba: Matsudo City Cultural Promotion Foundation, 2004); Tetsuo Matsumoto et. al. eds.,** *The World of Kenmochi: 4 – Chronology and Documents* **(Tokyo: Kawade Shobo Shinsha, Publishers, 1975); Valerie Fletcher,** *Isamu Noguchi: Master Sculptor* **(Scala Publishers, London, 2004).**

1841	**Herman Melville sets sail on the whaling ship** *Acushnet*, **departing from Fairhaven, Massachussetts, bound for Japan.**
1845	**Texas is annexed and the term "Manifest Destiny" is introduced in America.**
1853	**Commodore Matthew Perry sails an American armada of "Black Ships" into the bay of Edo, Japan.** **The** *Crystal Palace Exhibition*, **officially known as the** *Exhibition of the Industry of All Nations*, **opens in New York. It is the first international exposition held in the United States, and was inspired by London's** *Crystal Palace* **exhibition of 1851.** **Elisha Otis demonstrates his safe elevator at the American Institute Fair at the Crystal Palace in New York, under the management of P. T. Barnum.**
1854	**Treaty of Peace and Amity between the U.S. and Japan is signed, opening ports to U.S. whaling ships.**
1856	**Townsend Harris, the first U.S. Consul General, arrives in Shimoda, Japan.**
1858	**Treaty of Amity and Commerce between Japan and the U.S. is signed.** **Yukichi Fukuzawa founds a school in Tokyo that subsequently becomes Keio University.**
1861	**Civil War begins in the United States.**
1865	**Abraham Lincoln is assassinated. U.S. Civil War ends.**
1868	**The Meiji Restoration occurs and the process of modernization begins in Japan.**
1871	**Treaty of Amity between Japan and China is concluded.**
1873	**President Grant appoints John A. Bingham as U.S. Minister to Japan.**
1874	**Japan invades Taiwan.** **Leonie Gilmour (1874–1933), Isamu Noguchi's mother, is born in New York City.**
1876	**Japan recognizes Korean independence from China.** **Japan is invited to participate in the Centennial Exhibition held in Philadelphia's Fairmount Park. The exhibit celebrates the expansion of American industry and international commerce. Japan sends a twenty-five person commission to the exhibition and shows an array of porcelains, pottery, bronzes, lacquerware, furniture, screens, textiles, wood and ivory carvings, straw and bamboo items, and toys. Two complete traditional Japanese buildings are assembled for the exhibit. Ernest F. Fenollosa (1853–1908) visits the Centennial Exhibition and expresses critical acclaim for the Japanese exhibit.** **Thomas Edison founds his "invention factory" in Menlo Park, New Jersey.**

1877	Zoologist Edward Morse (1830–1925) arrives in Yokohama in search of coastal brachiopods. During his visit, he ends up collecting and cataloguing more than 5,000 Japanese ceramics.
1878	Ernest Fenollosa begins teaching at Kobu Daigakko (Tokyo University). Edison invents the incandescent light bulb. Westerners discover the tea ceremony with its emphasis on rustic utensils of simplicity, irregularity, and contrasts of rough and smooth surfaces. Fenollosa begins collecting Japanese ceramics.
1879	Former U.S. President Ulysses S. Grant visits Japan on a world tour. Tenshin Kakuzo Okakura, a student of Fenollosa, is credited along with Fenollosa for saving *Nihonga*, paintings produced by traditional Japanese techniques.
1881	British architect Josiah Conder builds the Tokyo Imperial Museum. Conder also becomes a professor at Kobu Daigakko (Tokyo University).
1886	Fenollosa sells his art collection to Charles Weld, a Boston physician, on the condition that it be donated to the Boston Museum of Fine Arts. The Statue of Liberty is dedicated in New York Harbor.
1887	The first Japanese fine-arts academy, Tokyo School of Fine Arts, is established with the help of Tenshin Kakuzo Okakura. Its emphasis is on Japanese traditional arts, epitomizing Japan's renewed interest in native culture in an effort to seek a modern Japanese form of artistic expression. Frank Lloyd Wright begins seven years of employment with the architecture firm Adler & Sullivan.
1889	The Eiffel Tower is completed and celebrated at the Paris Universal Exposition. Promulgation of the Constitution of the Empire of Japan.
1890	The Les XX exhibition of painters, designers and sculptors opens in Brussels. Writer Lafcadio Hearn (1850–1904) travels to Japan. Thomas Edison patents the motion picture camera and projection device. After living for many years in Japan, Fenollosa returns to Boston to become curator of the Oriental Department at the Museum of Fine Arts.
1891	Leonie Gilmour enters Bryn Mawr College. Yonejiro Noguchi (1875–1947), Isamu's father, enters Keio Academy in Tokyo.
1892	Leonie Gilmour wins a scholarship to study at the Sorbonne in Paris.

1893	**Fenollosa is asked to choose Japanese art for an exhibit at the Chicago World Columbian Exposition.** **Adolph von Hildenbrand publishes** *The Problem of Form in the Plastic Arts*. **Frank Lloyd Wright sees a replica of the Phoenix Hall (from Uji, Kyoto) at the World Columbian Exposition in Chicago.** **Yonejiro Noguchi sets sail for San Francisco from Japan.**
1894	**Sino-Japanese War begins over control of Korea.** **Western countries give up their extraterritorial rights in Japan.**
1895	**The Salon de l'Art Nouveau opens in Paris.** **Lumière brothers give first public motion picture showing.** **Wilhelm Röentgen discovers x-rays.**
1896	**Henry Ford completes his first experimental automobile.** **Five of Yonejiro Noguchi's English-language poems are published in** *The Lark*, **a San Francisco literary magazine.** **Bohemian Press publishes an English-language edition of Yonejiro Noguchi's first book of poems,** *Seen and Unseen, or Monologues of a Homeless Snail*.
1898	**Alexander Graham Bell visits Japan.**
1899	**Boxer Rebellion begins in China.** **Burton Holmes lectures on Japan at Columbia University, New York.** **Japan signs the Hague Agreement.**
1900	**Paris World's Fair opens in which the "modern style" triumphs in every medium, highlighting the evolution of Art Nouveau.** *Quantum Theory* **by Max Planck is published.** **First Zeppelin is developed.** **Pablo Picasso has his first exhibition in Barcelona, Spain.**
1901	**President McKinley is assassinated, and is succeeded by Theodore Roosevelt.** **Queen Victoria dies.** **Guglielmo Marconi transmits first signals by wireless telegraph.** **Frank Lloyd Wright presents his paper "The Art and Craft of the Machine" at Hull House in Chicago.** **Leonie Gilmour responds to an advertisement for an editor of works in English. The ad was placed by Yonejiro Noguchi.**

1902	**Anglo-Japanese Treaty is signed.** **Henry van de Velde, Belgian designer and architect known as the international artist of the Art Nouveau movement, is named head of the Grand-Ducal Saxon School of Arts and Crafts in Weimar, Germany, forerunner to the Bauhaus.** **Yonejiro Noguchi's** *The American Diary of a Japanese Girl* **is published in the United States.** **Yonejiro Noguchi travels from the U.S. to London.**
1903	**Construction of the Panama Canal begins.** **The Ford Motor Company is incorporated.** **Yonejiro Noguchi publishes** *From the Eastern Sea* **in London, New York, and Japan.** **Yonejiro Noguchi returns to the United States.**
1904	**Russo-Japanese War begins.** **Tenshin Kakuzo Okakura becomes the first head of the Asian art division of the Museum of Fine Arts, Boston, and publishes** *The Ideals of the East*. **Yonejiro Noguchi returns to Tokyo.** **Isamu Gilmour is born to Leonie Gilmour and Yonejiro Noguchi in Los Angeles, CA, on November 17.**
1905	**Japanese are victorious in the Russo-Japanese War.** **Albert Einstein publishes his paper "On the Electrodynamics of Moving Bodies."**
1906	**Tenshin Kakuzo Okakura publishes his influential** *Book of Tea* **that promotes the superiority of Asian art.**
1907	**New immigration law in the United States excludes Japanese workers.** **Leonie Gilmour takes her son Isamu to Tokyo to join his father, Yonejiro. Gilmour teaches English, and her students include the sons of Yakumo Koizumi.** **Yonejiro Noguchi marries Matsuko Takeda.**
1908	**Henry Ford produces the first Model T automobile.** **Frank Lloyd Wright designs the Robie House, Chicago, Illinois.** **The** *London Daily Mail* **sponsors the first exhibition of new "phases of home life" held at the Olympia Centre. Exhibits include construction, furniture, and decoration techniques, as well as an Arts and Crafts competition and a competition to design the "Ideal Home."** **Robert Henri, John Sloan, George Luks, William Glackens, Everett Shinn, and George Bellows form the "Ashcan School" of painters in Greenwich Village.**

1909 **Japanese Prime Minister Hirobumi Ito is assassinated in Ha'erbin, China.**

Bakelite, an early form of plastic, is patented.

Filippo Marinetti's *Futurist Manifesto* **is published.**

Yonejiro Noguchi publishes *Pilgrimage*, **poems in English dedicated to Leonie.**

1910 **Spanish architect Antonio Gaudí completes Casa Milá in Barcelona.**

A severe typhoon kills some 500 people in Japan and leaves approximately 400,000 homeless.

Japan annexes Korea.

Leonie Gilmour and Isamu move from Tokyo to the coastal town of Omori. Isamu enters Morimura Kindergarten.

1911 **The Chinese Revolution begins in Szechwan province. Six months later, a republic is declared in China, bringing an end to nearly three centuries of rule by the Manchu dynasty.**

First coast to coast U.S. flight is accomplished.

The Italo-Turkish War becomes the occasion for the first aerial bombardment in history.

1912 **Emperor Meiji dies and son Yoshihito (Emperor Taisho) is coronated.**

Sun Yat-sen is made provisional president of the Republic of China.

The *Titanic* **sinks in the North Atlantic.**

Japan and Russia reach an agreement on mutual spheres of influence in Manchuria and Mongolia.

Frank Lloyd Wright publishes a short treatise called *The Japanese Print: An Interpretation.*

French painter Marcel Duchamp completes *Nude Descending a Staircase.*

Wassily Kandinsky's *Concerning the Spiritual in Art* **is published.**

Picasso uses found objects to create collages, paintings, and sculpture.

Isamu and Leonie Gilmour move to the seaside town of Chigasaki in Kanagawa Prefecture, Japan. Isamu's half sister, Ailes, is born.

Isamu Kenmochi (1912–1971) is born on January 2 in Okubo village, Toyotama County, Tokyo, to Toshi and Tomi Kenmochi.

1913 **The Panama Canal opens.**

Henry Ford adopts a conveyor-belt technology for his factory.

California's Alien Land Act of 1913 prohibits those ineligible for citizenship, including Asian immigrants, from owning land.

Kazimir Malevich forms Suprematism movement in Russia.

The first home electric refrigerator is marketed in the United States.

The Armory Show introduces European modern art to New York City audiences.

Leonie Gilmour builds a house in Chigasaki and Isamu observes the craftsmanship of the carpenters. Isamu begins English school at St. Joseph's College, Yokohama.

1914–15	**Yuan Shih-k'ai establishes himself as virtual dictator of China.** **World War I begins.** **Japan declares war on Germany, siding with the Allied Forces and attacks Tsingtao, a German-controlled city in the Chinese province of Chiao Hsien.** **Frank Lloyd Wright builds Taliesin, a new home and architectural studio in Wisconsin.** **Nikakai (Nika Association, an artists' association) is established in Japan; founding members include Ryusaburo Umehara.** **Yonejiro Noguchi visits England.**
1916	**U.S. enters World War I.** **Noburo Ofuji invents the first animated film in Japan.** **Dada movement peaks in Zurich, spreading to Paris, Berlin, and New York.** **Frank Lloyd Wright is commissioned to design the Imperial Hotel in Tokyo (completed in 1922).**
1917	**October Revolution in Russia begins.** **Nippon Kogaku K.K., which will eventually become Nikon Camera Company, is established in Tokyo.** **Marcel Duchamp exhibits his readymades in New York.** **Piet Mondrian and Theo Van Doesburg's first issue of** *De Stijl* **is published.** **Isamu and his mother and sister move to Yokohama.**
1918	**World War I ends.** **Isamu Gilmour is sent alone to Rolling Prairie, Indiana, to attend Interlaken School, which closes a few months later. The founder of the school, Dr. Edward Rumely, places him with a Swedenborgian minister, Dr. Samuel Mack, in La Porte, Indiana, where he attends La Porte High School in 1919.** **Kenmochi enters Yodobashi elementary school in Tokyo.**
1919	**The Treaty of Versailles takes effect and is signed by leaders of the United States, Great Britain, France, and Italy—the agreement re-draws the map of Europe.** **Walter Gropius is appointed director of the former Grand Ducal School of Fine Arts of Saxony in Weimar. The school is officially fused with the School of Arts and Crafts, which was closed in 1915 and renamed Staatliches Bauhaus Weimar (State Bauhaus Weimar). The school is officially founded on April 1.** **Yonejiro Noguchi visits the United States but does not visit Isamu in Indiana.**
1920	**The League of Nations is formed.** **Marcel Breuer joins the Bauhaus and becomes one of six apprentices to join its new furniture workshop.** **Société Anonyme is founded in France.** **Kenmochi transfers to Shimoya elementary school.**

1921 **Walter Gropius and Adolf Meyer build the Haus Sommerfeld in Berlin in an Expressionist style. It is the first Bauhaus project involving an ideal of the unity of the arts in architecture.**

Gerrit Rietveld exhibits his furniture at the Bauhaus.

Kenmochi transfers to Takinogawa elementary school.

1922 **Mussolini forms fascist government in Italy.**

Ulysses **by James Joyce is published; U.S. post office burns copies.**

Marcel Breuer designs the De Stijl–influenced wood-slatted chair.

Vsevolod Meyerhold's *The Magnanimous Cuckold* **presents the first Constructivist stage sets by Liubov Popova.**

Albert Einstein visits Japan.

Tokyo Higher School of Arts and Technology is established (later renamed Chiba University).

Isamu Gilmour graduates from La Porte High School.

Dr. Edward Rumely arranges a summer apprenticeship for Isamu in Connecticut with sculptor Gutzon Borglum, and later Rumely raises money for Isamu to attend Columbia University, New York, in premedical studies.

1923 **The Great Kanto Earthquake hits the mainland Kanto area of Japan, creating an enormous need for new architecture and urban restructuring.**

In his writings, Kikutaro Shimoda, architect and rival of Frank Lloyd Wright, coins the term *teiken* **(crown-topped style).**

Frank Lloyd Wright develops the pre-cast Textile Block.

The Brooklyn Museum of Art exhibition *Contemporary Russian Painting and Sculpture* **opens.**

Isamu enters Columbia Medical School.

Kenmochi transfers to Shinno elementary school in Tokyo.

1924 **Congress passes a new and more restrictive law suspending Japanese immigration.**

J. Edgar Hoover is appointed director of the FBI.

Andre Breton's *Manifeste du Surréalisme* **is published.**

Leonie Gilmour moves to New York from Japan and encourages Isamu to take sculpture classes at the Leonardo da Vinci Art School in Manhattan. After three months studying with Onorio Ruotolo at the da Vinci School, and with a studio at 39 East 14th Street, Isamu has his first exhibition. He leaves Columbia Medical School, devoting himself to sculpture.

Taking his father's surname, Isamu Gilmour changes his name to Isamu Noguchi.

Noguchi moves to a studio at 127 University Place in Manhattan.

Kenmochi enters Azabu Middle School in Tokyo.

1925 **Physicists W. K. Heisenberg and Niels Bohr develop concept of quantum mechanics.**

The Bauhaus moves to new headquarters in Dessau, Germany, designed by Walter Gropius. Marcel Breuer accepts Gropius's invitation to return to the new Bauhaus as head of the furniture workshop. Breuer starts the development of the tubular steel Club Chair and the Wassily Chair No. B3, created for the painter Wassily Kandinsky.

Ford Motor Company buys the Stout Metal Airplane Company, which leads to the production of the Ford Trimotor aircraft in 1926.

Le Corbusier's *L'Art décoratif d'aujourd'hui* **is published, defining three different furniture types: type-needs, type-furniture, and human-limb objects.**

Alfred Steiglitz opens the Intimate Gallery in New York.

The first Surrealist group exhibition opens in Paris.

Noguchi begins work on portraiture, a practice that will financially sustain him for many years.

Noguchi has his first exhibition at the National Academy of Design, New York, and works in the theater for the first time by making masks for Michio Ito's performance in William Butler Yeats's *At the Hawk's Well.*

1926 **Germany is admitted into the League of Nations.**

Fascist youth organizations Hitlerjugend and the Italian Ballilla are founded.

Bell Telephone Laboratories transmits film by television.

The Kodak Company produces the first 16-millimeter movie film.

Alexander Calder creates his first wire-portrait sculptures in Paris.

Noguchi visits a Brancusi exhibition at the Brummer Gallery, New York.

1927 **Financial crash occurs in Japan.**

Daisetsu T. Suzuki publishes *Essay on Zen Buddhism.*

Charles Lindbergh completes the first transatlantic flight.

Marcel Breuer co-founds Standard-Möbel to manufacture and distribute his tubular steel furniture.

Eileen Gray designs the chair *At the Intersection between Luxury and Simplicity*, **inspired by deck chairs used on transatlantic steamers, manufactured in dark green or red lacquer using the Japanese** *Wajima* **technique.**

Charlotte Perriand begins her ten-year practice with Le Corbusier.

Noguchi is awarded a John Simon Guggenheim Fellowship to travel to Paris and the Far East.

Noguchi travels to Paris and works as Brancusi's studio assistant. Studies at the Académie de la Grande Chaumière, and meets other artists, including Alexander Calder and Stuart Davis.

1928 **The think tank of the modern movement (or International Style) in architecture, the Congrès Internationaux d'Architecture Moderne (CIAM), is founded in Switzerland.**

Andre Breton's *Surrealism and Painting* **is published.**

Le Corbusier begins experimenting with furniture design. Le Corbusier and Charlotte Perriand design the tubular steel LC2 Grand Comfort Armchair.

Picasso works with Julio González to create his first welded metal sculpture constructions.

Industrial Arts Research Institute (IARI) is established in Sendai, Japan.

Noguchi creates abstract sculpture and gouache drawings; moves to a studio at 11 rue Dedouvre, Gentilly, near Paris.

1929 **The U.S. Wall Street crash occurs.**

The Barcelona International Exposition opens.

The Museum of Modern Art opens in New York and exhibits works by Cezanne, Gauguin, Seurat, Van Gogh, and others.

Le Corbusier's Villa Savoye is designed.

R. Buckminster Fuller designs his Dymaxion House.

Ludwig Mies van der Rohe and Lily Reich design Chrome Leather Barcelona Chair.

Noguchi returns to New York and establishes a studio on the top floor of Carnegie Hall; meets R. Buckminster Fuller and Martha Graham.

Noguchi moves his studio to a hotel at Madison and 29th Street and is persuaded by Fuller to paint it silver. Noguchi continues to support himself with portrait-sculptures.

Kenmochi graduates from Azabu Middle School and enters the Woodworking Division of the Tokyo Higher School of Arts and Technology.

1930 **Japan signs the London Naval Treaty.**

Calder creates his first mobiles in Paris after visiting Mondrian's studio.

Alberto Giacometti produces his first Surrealist sculptures.

Noguchi travels to Chicago and Cambridge, MA, with R. Buckminster Fuller.

Noguchi has a successful exhibition of fifteen portrait-sculptures at Marie Sterner Gallery, New York.

Noguchi returns to Paris and travels to Moscow and Beijing. For seven months, he takes a studio at 18 Ta Yang Mao Hutung in Beijing and studies ink-brush techniques with Ch'i Pai-shih, creating a series of works known as the Peking Brush Drawings.

1931 **Japan invades Manchuria, setting off an undeclared war with China.**

Construction of Rockefeller Center in New York begins.

Empire State Building officially opens.

Junzo Sakakura begins a five-year assistantship with Le Corbusier in Paris.

Thomas Edison dies October 18, 1931, at the age of 84.

Noguchi travels to Japan but is not welcomed by his father. He briefly sets up studios in the Kodenma-cho district of Tokyo and the Higashiyama district of Kyoto, where he studies ceramics with Jinmatsu Uno.

Noguchi shows his ceramics in the *Nikaten Exhibition* **in Tokyo, then returns to New York in October.**

The Whitney Museum of American Art, New York, acquires Noguchi's portrait head of Ruth Parks.

Kenmochi spends the summer as a researcher at the Industrial Arts Research Institute (IARI) in Sendai.

1932 **Japan invades Shanghai, sparking the Shanghai War of 1932. China ultimately retreats. The League of Nations votes to censure Japan for its aggression in China.**

Franklin D. Roosevelt wins the U.S. presidency.

The first volume of *Kogei News* **(Industrial Art News) is issued by IARI.**

Noguchi exhibits his Peking Brush Drawings at Demotte Gallery and ceramics at John Becker Gallery in New York, and sets up a studio at 58 West 57th Street (The Sherwood Studios). Shortly thereafter he is evicted and takes another studio in a storefront at 446 East 76th Street, New York.

Noguchi's portrait head of Agnes Enters is acquired by The Museum of Modern Art, New York.

Kenmochi graduates from Tokyo Higher School of Arts and Technology, joins IARI in Sendai, and is assigned to the drafting, photographic printing, and carving division.

1933 **Japan and Germany resign from the League of Nations. Hitler assumes the government of Germany, and the National Socialists pressure the Bauhaus to close.**

Franklin D. Roosevelt establishes the Public Works of Art Project.

First full-length essay on Noguchi by Julien Levy is published.

Noguchi moves his studio to One West 67th Street (Hotel des Artistes), New York.

Leonie Gilmour dies in New York.

Noguchi makes models and drawings for Musical Weathervane, Playmountain, Monument to the Plough, and Monument to Ben Franklin; creates plaster models for Buckminister Fuller's Dymaxion Car; travels to London to exhibit his Peking Brush Drawings.

Kenmochi contributes writing to *Industrial Art News***; travels to Tokyo in April to establish the Tokyo branch of the IARI, then returns to Sendai in July and takes a summer course at the School of New Architecture and Industrial Arts in September.**

Bruno Taut visits the IARI exhibition at Mitsukoshi Department Store in Tokyo and criticizes prototypes designed at IARI. Taut is subsequently invited to advise IARI designers for five months, and Kenmochi serves as his assistant.

1934 **U.S.S.R. joins the League of Nations.**

Chinese Red Army makes its Long March.

American artist Mark Tobey spends a month studying calligraphy and painting at a Zen monastery near Kyoto.

Noguchi sets up a studio at 239 East 44th Street but spends the summer in Woodstock, New York.

A farewell party is given for Bruno Taut in March by staff at IARI in Sendai.

Kenmochi gives lectures on furniture design in Nagano, Japan.

Kenmochi and colleagues exhibit design prototypes in Ueno, Tokyo.

Kenmochi experiments with measuring the human form in a seated position using snow with Kappei Toyoguchi, a colleague at IARI.

1935 **The Nazis reject the Treaty of Versailles and establish the Nuremberg Laws.**

Italian forces invade Ethiopia.

Charlotte Perriand designs her first wood table.

Noguchi exhibits public projects and "social realist" works at Harriman Gallery, New York, and receives harsh reviews; creates first stage design for Martha Graham's *Frontier*.

Noguchi travels to California, where he designs a swimming pool (subsequently unrealized) at Richard Neutra's request for film director Joseph von Sternberg.

Late fall, Noguchi travels to Mexico City.

Kenmochi is appointed engineer of the Woodworking, Lacquerware, and Toy Division at IARI.

1936 **The Spanish Civil War begins.**

The Berlin Olympics open.

The Mingei aesthetic is promoted at the founding of the Japanese Folk Crafts Museum in Tokyo.

Marcel Breuer designs Molded Plywood Dining Table.

Henry Luce begins publication of *LIFE* **magazine**

Frank Lloyd Wright designs Fallingwater in Bear Run, Pennsylvania.

American Artists' Congress is organized to oppose fascism and publishes *Art Front* **in New York.**

Noguchi makes a social realist mural, *History of Mexico*, **a sixty-foot-long relief, at the Abelardo Rodríquez Market in Mexico City.**

Kenmochi gives lectures on design, manufacturing techniques, and processes for making wooden furniture for export, and is invited by local craftsmen and industries to speak in Niigata, Nagano, Tokyo, and Ibaraki. He begins to explore prototypes of woven crafts.

1937	**Junzo Sakakura builds his first East-West blend of architecture in the Japanese pavilion at the World Exposition in Paris and wins the Grand Prix. Eileen Gray displays a design for a cultural center in Le Corbusier's Pavilion at the Paris Exposition.**
	The Nazis organize the *Degenerate Art* **exhibition.**
	Roosevelt signs the U.S. Neutrality Act.
	The Vatican recognizes Franco's regime in Spain.
	Italy withdraws from the League of Nations.
	Japan launches full-scale invasion of China, bombing Nanking and Canton.
	Noguchi returns to his New York studio at 211 East 49th Street and designs his first manufactured object, Radio Nurse, a baby monitor in Bakelite for the Zenith Radio Company.
	Kenmochi marries Kyoko Maiya.
1938	**President Roosevelt appeals for peace directly to Hitler and Czechoslovak President Benes.**
	The Munich Pact is signed by Daladier, Chamberlain, Mussolini, and Hitler. Germany renounces any further territorial claims against its neighbors.
	Japan begins regular bombing of Chungking and is condemned by the League of Nations for its use of poison gas in China.
	The U.S. State Department rejects Tokyo's claim of a "New Order" in Asia.
	The U.S. House Committee on Un-American Activities is formed to investigate disloyalty and suspected subversive organizations.
	David Smith has his first solo exhibit of welded iron sculptures and drawings at Marian Willard's East River Gallery, New York.
	Noguchi designs his first fountain for the Ford Motor Company Building, New York World's Fair.
	Noguchi is offered a commission to design a wall plaque for the Associated Press Building, Rockefeller Center, New York.
	Kenmochi gives lectures on woven and industrial products in Ibaraki and Iwate. His first son is born.
1939	**Nazi Germany invades Poland, World War II begins.**
	The first jet-powered aircraft, the Heinkel 178, makes its first flight.
	The World's Fair opens in New York.
	Buckminster Fuller's Dymaxion Bathroom is exhibited at The Museum of Modern Art, New York.
	The Museum of Non-Objective Painting (later renamed the Solomon R. Guggenheim Museum) opens in New York.
	Mies van der Rohe is commissioned to design the campus of the Illinois Institute of Technology.
	Noguchi travels to Hawaii for the Dole Company and designs his first playground equipment (ultimately unrealized) for Ala Moana Park in Honolulu; returns to New York and moves to 52 West 10th Street, and designs his first table for A. Conger Goodyear, president of The Museum of Modern Art.
	Kenmochi exhibits his celebrated Fruit Basket; lectures on wooden and bamboo products for export, giving technological advice to manufacturers in Nagoya, Niigata, Yamagata, and Hyogo.

1940 The Nazis invade France, Belgium, Luxembourg, and the Netherlands; Italy declares war on the Allies; the Germans enter Paris; the Japanese occupy Indochina. The Battle of Britain begins.

D. T. Suzuki's *Zen Buddhism and Its Influence on Japanese Culture* is published in Kyoto.

Charlotte Perriand visits IARI in Japan.

The Lascaux caves are discovered.

The U.S. Nationality Act is enacted, which declares that American citizens born of foreign parents can lose their citizenship if they reside abroad for more than six months.

Noguchi designs the set for Martha Graham's *El Penitente*.

IARI moves to Tokyo. Kenmochi moves to Tokyo

1941 The Japanese attack Pearl Harbor.

Italy and Hungary invade Yugoslavia; Russo-Japanese neutrality pact is signed; Japan invades Malaya.

Hitler breaks nonaggression pact with Stalin and attacks Russia.

Britain declares war on Finland, Hungary, and Romania; Britain and the United States declare war on Japan; the United States declares war on Germany; the Manhattan Project begins.

The first computer controlled by software, Konrad Zuse's Z3, is invented.

Antonin Raymond designs Carrera House in Montauk Point, New York.

Motorola Company manufactures a two-way AM police radio.

Victor Hasselblad of Sweden forms the Hasselblad camera company.

Charlotte Perriand exhibits her furniture at Takashimaya Department Store in Tokyo and Osaka.

Noguchi constructs a model for Contoured Playground (unrealized) for Central Park; drives to California with Arshile Gorky and friends; creates a laminated table made from primavera wood (accessioned to The Museum of Modern Art Design Collection in 1958).

Kenmochi's second son is born.

1942 U.S. citizens of Japanese descent on the West Coast are sent to internment camps.

Breton and Duchamp present *First Papers of Surrealism* exhibition in New York.

Peggy Guggenheim opens her Art of This Century Gallery in New York.

Perriand is forced to leave Japan as an "undesirable alien," but is trapped by a naval blockade and forced to spend the rest of the war in Vietnam.

Noguchi travels to Washington, D.C., attempting to mitigate the hardships of Japanese-Americans.

In spring, the San Francisco Museum of Modern Art mounts an exhibition of Noguchi's work.

Noguchi voluntarily enters a relocation center in Poston, Arizona, but leaves in November, returning to New York; establishes a new studio at 33 MacDougal Alley.

1943 **Germany retreats from Stalingrad.**

Allied Forces invade Italy and Italy surrenders.

Frank Lloyd Wright is offered the commission to design the Solomon R. Guggenheim Museum in New York.

Noguchi creates his first illuminated sculptures, called "Lunars."

Kikaku Kagu **(Standardized Furniture) is published.**

Kenmochi and others at IARI conduct research for the war effort on the wooden airplane.

1944 **Rome and Paris are liberated.**

Franklin D. Roosevelt elected to a fourth term as U.S. president.

Laszlo Moholy-Nagy founds the Institute of Design in Chicago.

Industrial Art News **suspends publication.**

Noguchi designs three dance sets for Martha Graham, *Appalachian Spring*, *Herodiade*, **and** *Imagined Wing*, **and a set and costumes for Ruth Page's** *The Bells.*

Noguchi designs biomorphic Glass-Topped Coffee Table and Rudder Dining/Coffee Table.

Noguchi designs chess set and chess table for *The Imagery of Chess* **exhibition at the Julien Levy Gallery, New York.**

Noguchi designs furniture and stair, wall, and ceiling reliefs for the reception room at the Time and Life Building, New York.

Kenmochi works as an engineer for airplane munitions.

Kenmochi's first daughter is born.

1945 **The Yalta Conference is convened; President Roosevelt dies; Germany surrenders.**

The United Nations charter signed.

The IARI headquarters burns in the fire-bombing of Tokyo.

Atomic bombs are dropped on Hiroshima and Nagasaki.

Japan surrenders, and World War II ends.

The Japanese Exclusion Act is repealed.

Mark Rothko exhibits at the Art of This Century Gallery, New York.

Knoll Associates, Inc., produces Noguchi's Three-Legged Cylinder Lamp (until 1954).

Noguchi proposes a project (unrealized) for Jefferson Memorial Park in St. Louis, MO; travels to Ohio to visit the Great Serpent Mound; designs dance sets for Erick Hawkins's *John Brown and God's Angry Men*.

Kenmochi transfers to the Tohoku branch of IARI.

1946 **U.S. conducts atomic explosion tests on Bikini Atoll.**

First U.N. General Assembly convenes in Paris.

Charlotte Perriand returns to France and revives her career as an independent designer through her collaboration with Jean Prouvé.

Writing in *The New Yorker*, **Robert M. Coates applies the term "abstract expressionism" to certain New York painters.**

Ruth Benedict publishes *The Chrysanthemum and the Sword: Patterns of Japanese Culture* **in Boston.**

Noguchi exhibits in *Fourteen Americans* **at The Museum of Modern Art; Thomas B. Hess publishes "An Art News Contemporary Contour: Isamu Noguchi" in** *Art News*.

Kenmochi designs furniture for the residences of the U.S. Occupation Forces. Kenmochi and others research the production of standardized furniture at IARI.

Kenmochi's third son is born.

1947 **Japan signs the MacArthur Peace Treaty with the United States.**

The House Un-American Activities Committee holds nine days of hearings into alleged Communist propaganda in the Hollywood motion-picture industry.

Yonejiro Noguchi dies in Tokyo.

Noguchi produces a model for *Sculpture to Be Seen from Mars* **(unrealized). Creates lunar ceiling for the American Stove Company headquarters in St. Louis, MO. Herman Miller begins production of several of Noguchi's furniture designs.**

Noguchi designs two sets for Martha Graham, *Errand into the Maze* **and** *Night Journey*, **and sets for** *Stephen Acrobat* **by Erick Hawkins and** *The Four Seasons* **by Merce Cunningham and John Cage.**

Two Noguchi works are included in the *Exposition Internationale du Surréalisme*, **Galerie Maeght, Paris.**

Kenmochi and others at IARI research wood and bamboo products for export and the mass-production of wood products.

1948 **Mahatma Gandhi is assassinated.**

The United Nations is established. The U.S. Congress approves the Marshall Plan.

Jackson Pollock shows his drip paintings at Betty Parsons Gallery, New York.

Painter Arshile Gorky commits suicide in Connecticut.

Architect Antonin Raymond designs the Reader's Digest Building in Tokyo.

Noguchi's designs this year include: a York fossil marble table and a Rosewood table for Samuel Dretzin; a park at Raj-gat and Memorial for Gandhi (unrealized); sets for Martha Graham's *Diversion of Angels* **and Yuriko Anemiya's** *Tale of Seizure*, **and a set and costumes for Balanchine's** *Orpheus*; **a laminated beechwood table, chairs, and irons for William A. M. Burden, and the Marble-Topped Coffee Table for retail sale at Bloomingdale's; nine designs for ashtrays (prototypes for a magazine article), and designs a lighted lunar wall for the SS** *Argentina* **stairwell, and a free-form couch and ottoman produced by Herman Miller.**

Kenmochi and others at IARI work on prototypes of wood and bamboo products for export. Kenmochi researches and lectures on a technique for applying high-frequency heat to woodworks. IARI headquarters moves to the Ota Ward in Tokyo.

1949 **The Communist People's Republic of China is established. Russia tests an atomic bomb.**

In London, D. T. Suzuki publishes an English version of *Introduction to Zen Buddhism*, **with a preface by Carl Jung.**

Kenzo Tange begins designing the Peace Center in Hiroshima.

Noguchi exhibits recent sculptures at Charles Egan Gallery, New York, and is awarded a Bollingen Foundation grant to travel around the world to gather research for *The Study of Leisure*. **In May he begins travel to Europe, the Middle and Far East, and Southeast Asia.**

Kenmochi and others at IARI continue to explore the manufacture of high-quality mass-produced wood and bamboo products and the use of high-frequency heat.

Kenmochi gives a lecture on product finishes in Oita.

1950 **North Korea invades South Korea. The U.S. sends troops to South Korea and the Korean War begins.**

The Temple of the Golden Pavilion burns in Kyoto, Japan.

Harry Bertoia designs the Wire Chair.

David Smith produces large-scale welded metal sculptures.

In May Noguchi travels to Japan after almost twenty years and meets Saburo Hasegawa, who guides him through the old temples and gardens of Japan; rereads D. T. Suzuki's *Zen and Its Influence on Japanese Culture*.

Yoshiro Taniguchi asks Noguchi to design a faculty room and garden for Keio University in Tokyo.

Noguchi and Kenmochi meet, and Kenmochi offers Noguchi a studio at the IARI.

Noguchi exhibits his ceramics, furniture, and sculptures for the Keio University project, and two models of *Bell Tower for Hiroshima* **at the Mitsukoshi Department Store in Tokyo.**

Kenmochi is appointed chief of the research division at IARI; designs prototype for the Bamboo Basket Chair with Noguchi.

Noguchi returns to New York and meets movie actress Yoshiko Yamaguchi at a fashion show at the Brooklyn Museum.

Kenmochi and others at IARI research the molding fine plywood with high-frequency heat, and explore the qualities of synthetic resin coating. Kenmochi gives a lecture in Kanagawa on seasoning wood; joins the newly established architectural division of Shin-Seisaku Kyokai (Association of New Works); and exhibits design works.

1951 **Japan regains status as an independent country.**

The San Francisco Treaty of Peace, a treaty of mutual cooperation and security, is signed between the U.S. and Japan.

In Japan, the Bokujin-kai (Ink Human Society) is founded to transform traditional calligraphy into modern Expressionist painting.

R. Buckminster Fuller's Geodesic Dome is exhibited at The Museum of Modern Art, New York.

Louis I. Kahn wins his first major commission, an extension to the Yale University Art Gallery.

Junzo Sakakura designs the Museum of Modern Art in Kamakura, Japan.

Noguchi returns to Japan with Yoshiko Yamaguchi; visits Gifu, and designs his first lantern, called Akari, the Japanese word for light.

Noguchi designs two bridge railings for Hiroshima Peace Park (completed 1952), and creates a garden for the Reader's Digest Building in Tokyo.

Noguchi exhibits marble sculptures in the first São Paulo Biennial in Brazil.

Noguchi accompanies Yoshiko Yamaguchi to Hollywood and returns to New York to work on a plaza for the Lever House designed by Gordon Bunshaft of Skidmore, Owings & Merrill.

The first exhibition organized by IARI, *Design and Technology*, opens at Mitsukoshi Department Store in Tokyo.

Kenmochi is appointed chief of engineering in the Engineering Division of IARI, and gives lectures on new furniture design and the characteristics of coating and its various uses.

1952 The Allied occupation of Japan ends.

The U.S. tests first hydrogen bomb.

Alien Land Act of 1913 is ruled unconstitutional by California's Supreme Court.

Le Corbusier is asked to design buildings in Chandigarh, India, with Iannis Xenakis.

Frank Lloyd Wright gives a lecture in New York titled "The Destruction of the Box" inspired by Tenshin Kakuzo Okakura.

John Cage and Robert Rauschenberg collaborate on *Untitled Event* at Black Mountain College. Cage composes using the I-Ching, and also premieres *4'33"*, a Zen-inspired performance.

D. T. Suzuki gives his influential lectures on Zen philosophy and culture, which continued until 1957, at Columbia University, New York.

Harold Rosenberg coins the term "action painting."

Noguchi is asked by Kenzo Tange and the Mayor of Hiroshima to design a cenotaph for Hiroshima City. The design is ultimately rejected.

Noguchi meets Rosanjin Kitaoji and Kaneshige Toyo; designs more Akari Light Sculptures in Gifu.

Noguchi designs Four-Legged Square Lamp, and Knoll Associates, Inc., produces it, but it is overrun by cheaper imitations.

In September, Noguchi has a large ceramic exhibition at the Museum of Modern Art, Kamakura, where Akari are also shown.

Noguchi marries Yoshiko (Shirley) Yamaguchi.

Noguchi designs this year include a plastic wall lamp, an illuminated mobile, a ceiling fixture, and two other lamps, produced for Lightolier and Limited Editions.

Kenmochi helps to establish the Association of Industrial Designers, as well as the Competition of Industrial Design in Japan, organized by Mainichi Newspaper Corporation, for which he serves as a juror. Kenmochi is appointed manager of the Design Division at the IARI; from April 30 to November 29 he travels to the United States (San Francisco, Los Angeles, Minneapolis, Chicago, Detroit, and New York) and meets with Charles and Ray Eames, Mies van der Rohe, Philip Johnson, Raymond Loewy, Edgar Kaufmann, Jr., and George Nakashima. On August 21 he is interviewed by the *New York World-Telegram* and the *Sun*, and discusses "Japanese Modern" with George Nelson in September.

1953 **The Korean War ends. Josef Stalin dies. Dwight Eisenhower is elected U.S. president.**

Betty Parsons exhibits her first show of Japanese artists.

Marcel Breuer begins design of UNESCO headquarters in Paris with Pier Luigi Nervi and Bernard Zehrfuss.

Antonin and Noemi Raymond design Hayata House and Cunningham House in Tokyo; both are expressions of their desire to reform postwar Japanese house design.

Architect Togo Murano completes the World Peace Memorial Cathedral in Hiroshima Jôzsef Révai's "Tradition and Modernism in Architecture" is translated into Japanese and printed in the October issue of *Bijutsu Hihyo*.

Noguchi returns to New York in January and reunites with Yamaguchi in Paris in July. They travel together for several months, and he returns to New York in October, and works to reverse the U.S. Government's refusal to renew Yamaguchi's visa, denied because of her associations with Communists.

Bijutsu Shuppansha, Tokyo, publishes *Isamu Noguchi*, **with essays by Shuzo Takiguchi, Saburo Hasegawa, and Noguchi. Lever House commission is cancelled.**

Kenmochi designs a wardrobe made of paulownia and metal; creates furniture design for a living-room display at the International Trade Fair in Canada; designs furniture for a living-room display at the *17th Shin-Seisaku Exhibition* **at the Tokyo Metropolitan Museum of Art; designs a display for the** *National Electrical Appliances* **show at Hankyu Department Store, Osaka; continues research on Japanese Modern style with others at IARI. Travels to the U.S. to attend the Aspen International Design Conference in Colorado.**

Kenmochi is appointed to the Japan Industrial Designers Association Committee.

1954 **The U.S. Senate votes to censure Senator Joseph McCarthy for his pattern of behavior in connection with anti-Communist activities.**

The Museum of Modern Art, New York, presents a series of exhibitions focused on Japanese culture; the first is the Japanese Exhibition House.

Ad Reinhardt publishes a series of articles on Asian art in *Art News*. **The Gutai (Concrete) group is founded in Ashiya, Hyogo Prefecture. Akira Kurosawa directs** *The Seven Samurai*.

Peter Voulkos creates original clay sculptures inspired by Japanese ceramics.

Noguchi has solo exhibition of ceramics at the Stable Gallery, New York. Knoll Associates manufactures Noguchi's Table and Stool.

Walter Gropius visits the IARI.

Kenmochi serves on a committee of Japan UNESCO (until1971); designs combination furniture and bamboo chair. A living room set of low-cost furniture is shown at the *Shin-Seisaku* **exhibition at Matsuya Department Store in Tokyo. His low-cost furniture is also shown at the** *18th Shin-Seisaku Exhibition* **at the Tokyo Metropolitan Museum of Art.**

Design and Technology **exhibition is organized by the IARI at the Mitsukoshi Department Store.**

"Japonica Controversy" begins.

Kuwasawa Design Laboratory is established and Kenmochi is appointed a lecturer on "Structure."

1955 The Museum of Modern Art presents an exhibition of ceramics by Noguchi's mentor, Kitaoji Rosanjin.

Charlotte Perriand exhibits her furniture at Takashimaya Department Store in Tokyo.

Sony Corporation markets the first transistor radio.

First Gutai art exhibition is held in Tokyo.

D. T. Suzuki publishes *Studies in Zen*.

Noguchi designs a house for two painters in Carmel, CA (unrealized). His first Akari exhibition in New York opens at Bonniers. He designs sets for the Royal Shakespeare Company's production of *King Lear*, and sets for Martha Graham's *Seraphic Dialogue*, and creates iron sculptures in Gifu, Japan.

Japanese Modern Furnishing model room is planned by the IARI at the International Industrial Design: Housing, Furniture, and Crafts Expo in Helsingborg, Sweden.

Kenmochi's combination furniture is shown at the *19th Shin-Seisaku Exhibition* at the Tokyo Metropolitan Museum of Art; designs a dining chair using bamboo and wood at the IARI; serves as a chairman for seminars "Design in the United States"; resigns from the IARI and establishes his own firm, Kenmochi Design Associates (KDA).

Writer and architect Bernard Rudofsky visits Japan and KDA.

1956 Japan joins the United Nations.

Egypt nationalizes the Suez Canal, setting off an invasion by England, France, and Israel.

The U.S. tests the first aerial H-bomb on Bikini Atoll in the Pacific.

American painters Sam Francis, Ad Reinhardt, and Mark Tobey travel to Japan.

Charles and Ray Eames design the Bentwood Lounge Chair and Ottoman.

Martha Jackson Gallery in New York hosts a Gutai exhibition.

Noguchi begins design of UNESCO headquarters gardens in Paris (Marcel Breuer architect) and Connecticut General Life Insurance Company, as well as a lobby waterfall wall and ceiling for 666 Fifth Avenue, New York.

Noguchi and Yoshiko Yamaguchi divorce.

Kenmochi helps establish Japanese Interior Designer Association.

Kenmochi establishes Living Art Showroom, displaying items selected by his firm.

1957 Alan Watts publishes *The Way of Zen* in New York.

Yukio Mishima, author and playwright, noted for his nihilistic postwar writing, wins the Yomiuri Prize for *The Temple of the Golden Pavilion* from Yomiuri Newspaper Co.

Charlotte Perriand designs the League of Nations building for the United Nations in Geneva.

Noguchi returns to Japan to search for stones for UNESCO gardens; designs Prismatic Table for the Alcoa Aluminum Company.

Kenmochi designs interior and furniture for the Shiseido Beauty Salon, Tokyo; serves on the selection committee for Japanese craftworks to be represented in the Japanese Pavilion at the '58 World Expo in Brussels, Belgium; meets with Charles and Ray Eames on their visit to Japan.

1958 **The Museum of Modern Art exhibits** *Abstract Japanese Calligraphy.*

Noguchi creates sheet-metal sculptures at the lighting factory of Edison Pierce in Long Island City, New York, with the assistance of Shoji Sadao. Creates set designs for Martha Graham's *Clytemnestra* **and** *Embattled Garden.*

Kenmochi Design Associates designs the governor's desk and sideboard at the Kagawa Prefecture Government Office (Kenzo Tange, architect). KDA designs Flexible Table, Tea Table, and Stacking Stool 202, displayed at the *Apartment Interior* **exhibition at Ginza Matsuya Department Store, Tokyo. Kenmochi wins a gold medal for his interior design of the Japanese Pavilion (Kunio Maekawa, architect) at the World Expo in Brussels. Serves on a committee for the Japanese Interior Designer Association, and is commissioned to design a showroom for the Mitsubishi Corporation.**

1959 **Junzo Sakakura designs Hajima City Hall, Japan.**

Louis I. Kahn designs the Salk Institute in La Jolla, California.

Allan Kaprow presents *18 Happenings in Six Parts* **at the Reuben Gallery, New York.**

Noguchi begins a series of balsa wood sculptures; exhibits marble, granite, and iron sculptures at the Stable Gallery, New York.

Kenmochi becomes president of the Japanese Designer Association Committee (until 1960) and serves on the design committee for the Tokyo Olympics (until 1965); serves on the design promotion council committee organized by the Ministry of International Trades and Industry (until 1971); becomes a lecturer at Tama Art University (until 1964).

KDA designs Rattan Round Chair C-315-O, fabricated by Yamakawa Rattan (currently C-3170; still in production by Y.M.K. Co., Ltd.).

Stacking Stool 202 distribution begins (still in production by Akita Mokko Co., Ltd.); KDA designs a booth for Japanese music instruments at the International Trade Fair in Los Angeles.

1960 **The Vietnam War begins.**

The Treaty of Mutual Cooperation and Security between Japan and the U.S. is signed.

Kenzo Tange designs Kurashiki City Hall in Japan.

Horizon of Japanese Design **exhibition is held at Toyoko Department Store, Tokyo.**

Noguchi begins design for the Sunken Garden at the Beinecke Rare Book and Manuscript Library at Yale University and The Billy Rose Sculpture Garden at Israel Museum, Jerusalem; creates set designs for Martha Graham's *Acrobats of God* and *Alcestis*; **exhibits at Documenta II, Kassel, Germany.**

Kenmochi withdraws from Japan Industrial Designers Association; serves as a panelist at the World Design Conference in Tokyo. KDA designs this year include: Armchair 207 (currently reproduced by Akita Mokko Co., Ltd.); furniture for Ohtone Country Club (Toshio Yamashita, architect); Rattan Round Chair C-315-S (currently C-3150; still in production); Rattan Round Chair C-315-E, fabricated by Yamakawa Rattan (currently C-3160; still in production by Y.M.K. Co., Ltd.); interior of the Hotel New Japan, including designs for a hanging lamp and a lounge table.

Exhibition Design: Isamu Kenmochi and Riki Watanabe **opens at Oriental Nakamura Department Store, Nagoya, Japan.**

1961 **The Berlin Wall is built.**

John F. Kennedy is elected U.S. President.

Olympic area construction begins in Japan, designed by Kenzo Tange, along with Tange's Ichinomiya Rowhouses.

Kunio Maekawa designs the Metropolitan Festival Hall in Tokyo.

Noguchi designs the Sunken Garden for Chase Manhattan Bank Plaza, New York; establishes studio and residence in a former factory building in Long Island City, Queens, New York; exhibits stainless-steel sculptures at the Cordier and Warren Gallery; is commissioned to create a garden for Chase Manhattan Bank Plaza; collaborates with Louis I. Kahn on Riverside Drive Playground (unrealized).

KDA designs Table, Stool, and Kashiwado Chair for the Atami Garden Hotel (Kenzo Tange, architect; fabricated by Tendo Co., Ltd.), interior and furniture for the Totsuka Country Club (Kenzo Tange, architect); interior and furniture for the IBM Chidori-cho Factory; and a display for the *Good Design* **exhibition at Matsuya Department Store.**

1962 **Astronaut John Glenn orbits the earth three times.**

The TWA Terminal is designed by Eero Saarinen for JFK Airport in New York.

Andy Warhol paints *32 Campbell's Soup Cans.*

Noguchi completes Mississippi Fountain for John Hancock Insurance Company Building (now Kaig and Besthoff Building), New Orleans; designs set for Martha Graham's *Phaedra*; **travels to Rome as an artist-in-residence at the American Academy; works in the marble quarries of the firm Henraux in Querceta, Italy. Akari exhibition opens at Bonniers in New York.**

KDA designs interior and furniture for the New Orient Express. Kenmochi serves on the selection committee for the *Good Design* **exhibition.**

1963 **Nuclear test-ban treaty signed.**

Martin Luther King, Jr., delivers "I Have a Dream" speech.

U.S. President John F. Kennedy is assassinated.

The Kobe Port Tower in Japan, a hyperboloid structure, is completed.

KDA designs interior and furniture for the Daiichi Hotel and the Keyaki Grill at the Tokyo Hilton Hotel, and wins the Ninth Industrial Design Award from Mainichi Newspaper Corporation.

1964 **People's Republic of China tests an atomic bomb.**

Robert Rauschenberg is awarded first prize at the Venice Biennale.

U.S. Tonkin Resolution permits presidential action in Vietnam.

U.S. Civil Rights Act prohibits discrimination on the basis of color, race, religion, or national origin.

Cosmic microwave background radiation is discovered.

The World's Fair opens in Flushing Meadows Park, New York.

Kenzo Tange's Tokyo Olympic Arena is completed.

Architecture Without Architects **by Bernard Rudofsky is published.**

Hiroshi Teshigahara directs *Woman in the Dunes.*

Noguchi creates gardens for IBM headquarters in Armonk, New York. Designs a tomb for John F. Kennedy (unrealized), and a set for Martha Graham's *Phaedra.*

KDA designs include: interior and furniture for Keio Department Store; Table T-209, fabricated by Yamakawa Rattan (still in production by Y.M.K Co., Ltd); and Utensils "SASA." Rattan Round Chair C-315-E is selected for The Museum of Modern Art Design Collection; Kenmochi joins the Japan Institute of Architects, and takes a professorship at Tama Art School (until 1968).

1965 **The first American combat troops land in Vietnam.**

The U.S. Northeast Blackout leaves approximately 25 million people in the U.S. and Canada without power for twelve hours.

John Lindsay is elected mayor of New York City.

Commissioned by financier Matsutaro Shoriki, R. Buckminister Fuller creates a visionary design for Tetrahedron City (unrealized) for one million inhabitants in Japan.

Noguchi begins first playground design for Kodomo No Kuni (children's playground) near Tokyo for Children's Year Festival in 1966; exhibits carved stone sculptures at the Cordier and Ekstrom Gallery, New York.

KDA designs Circular Bench and Bench for the Kagawa Prefecture Gymnasium (Kenzo Tange, architect), fabricated by Tendo Co., Ltd.

September 14 to November 13, Kenmochi travels to Switzerland, Germany, the United Kingdom, the United States, Mexico, and Guatemala.

1966 **Mao Zedong launches the Cultural Revolution in China.**

The Museum of Modern Art, New York, mounts a retrospective exhibition of Louis I. Kahn's work.

Commissioned by Matsutaro Shoriki, R. Buckminster Fuller creates a design for a two-mile-high tower.

Construction begins on the World Trade Center in New York, designed by architect Minoru Yamasaki.

Robert Venturi publishes his influential *Complexity and Contradiction in Architecture.*

Marcel Breuer designs a building for the Whitney Museum of American Art, New York.

Noguchi designs a set for Martha Graham's *Cortege of Eagles*, **his last collaboration with Graham; completes the Billy Rose Sculpture Garden for The Israel Museum, Jerusalem; creates a monumental painted metal sculpture for the Museum of Modern Art in Tokyo; establishes Akari Foundation in New York; and receives the commission for his** *Black Sun* **sculpture from the Seattle Art Museum, which he begins in Mure, Japan.**

KDA designs this year include: the Easy Chair F; Big Ashtray; Trash Bin; carpet and other furniture for the Kyoto International Conference Hall (Sachio Otani, architect); a daybed fabricated at Yamaguchi Mokuzai Kogei.

Kenmochi serves on the selection committee for the 1966 *Good Design* **exhibition.**

1967 **Israel declares victory in the Six-Day War.**

Large-scale anti-war protests are held across the United States. In New York, the protest begins in Central Park with a march to the United Nations building led by the Rev. Martin Luther King, Jr.

Race riots break out in Newark, NJ, and Detroit, MI.

Louis I. Kahn designs the Kimbell Art Museum in Fort Worth, TX.

Noguchi exhibits metal and stone sculptures at the Cordier and Ekstrom Gallery, New York.

KDA designs Love Seat (Settee), fabricated by Yamakawa Rattan (still in production by Y.M.K. Co., Ltd.), and the showroom at Kenmochi Design Associates.

Kenmochi travels to the United States and Canada, and visits the Montreal Expo; meets with Henry Dreyfuss, George Nelson, and Charles and Ray Eames; serves on the design council for the Winter Olympics in Sapporo; and becomes a lecturer at the Department of Architecture, Science, and Engineering at Waseda University.

1968 **The USS** *Pueblo*, **an American intelligence-gathering vessel, is captured by North Korea.**

Japanese GNP ranks second in the world.

Yasunari Kawabata wins the Noble Prize for Literature.

Martin Luther King, Jr., and Robert F. Kennedy are assassinated.

Richard Nixon wins the U.S. presidency.

Philippe Starck founds his first design firm, specializing in inflatable objects.

Noguchi begins a series of stone table sculptures and a marble series; designs *Red Cube* **for lower Manhattan, and the red Octetra play sculpture for Spoleto, Italy; exhibits Akari at** *Noguchi: Shapes of Light* **at the Cordier and Ekstrom Gallery; and publishes his autobiography,** *A Sculptor's World*. **In April a Noguchi retrospective opens at the Whitney Museum of American Art, New York, and a solo exhibition of recent sculptures opens at the Gimpel and Hanover Gallery in London and Zurich.**

KDA designs the first-class lounge for Japan Airlines B-747, and interiors and furniture for Tokyo Kaikan (Yoshinobu Ashihara, architect).

Kenmochi travels to Taiwan in August and to China in September.

1969 **Apollo 11 manned space mission lands on the moon.**

Louis I. Kahn begins building the Yale Center for British Art, Yale University.

The Japan Industrial Design Promotion Organization (JIDPO) is established.

The withdrawal of U.S. forces from Vietnam begins in July.

Noguchi creates *Skyviewing Sculpture* **for Western Washington University, and installs** *Black Sun* **in Seattle.**

Noguchi establishes a studio in the village of Mure-cho in Shikoku, Japan.

KDA designs the interior of the Yamagata-ya Noriten Store and the Easy Chair SM 7051, made of transparent acrylic and metal and fabricated by Tendo Co., Ltd.

In January and February, Kenmochi travels to the United Kingdom, France, Sweden, Finland, Germany, and the USSR; visits the International Design Expo in Paris and the International Lamp Show in Göteborg, Sweden.

From April 29 to May 1, Kenmochi travels to Hong Kong and Macao.

Beginning in May, Kenmochi serves as an adviser for the display committee of the 1970 World Expoin Osaka.

In June, Kenmochi serves on the selection committee for the Yamagiwa Lighting Design Competition.

The IARI is reorganized under the former Agency of Industrial Science and Technology (today's National Institute of Advanced Industrial Science and Technology).

1970 U.S. forces enter Cambodia but ground troops pull out several months later. The United States repeals the Gulf of Tonkin Resolution.

Writer Yukio Mishima commits suicide.

Lithium, a drug used to treat manic depression, is approved by the FDA.

Osumi, Japan's first satellite, is launched on a Lamba-4 rocket.

A Chinese nuclear test is conducted in Lop Nor.

The Nuclear Non-Proliferation Treaty goes into effect, ratified by forty-three nations.

The Expo '70 World's Fair opens in Osaka, Japan.

Dr. Timothy Leary, proponent of LSD, is sentenced to ten years in jail.

China's first satellite is launched into orbit.

The North Tower of the World Trade Center is completed at 1,368 feet, making it the tallest building in the world.

R. Buckminister Fuller receives the AIA Gold Medal.

Artists Barnett Newman and Eva Hesse die.

Noguchi designs nine fountains for Expo '70 in Osaka, Japan.

Akari are introduced at Bloomingdale's Department Store in New York.

Noguchi has a solo exhibition at the Cordier and Ekstrom Gallery, New York.

KDA designs include: interior of Kikkoman Soy Sauce Restaurant at Expo '70; interiors and furniture for the Time-Life International Building; Bench SF-70105 for Expo '70 in Osaka; and displays for the *Expo '70 Street Furniture Design* exhibition at Matsuya Department Store in Tokyo.

KDA exhibits chairs at the *34th Shin-seisaku Exhibition* at Tokyo Metropolitan Museum of Art.

A monument to the founding of Industrial Arts Research Institute is installed in Sendai, Japan.

Kenmochi wins the Mainichi Newspaper Corporation's 15th Industrial Design Award for *Expo '70 Street Furniture Design*. Survives a serious car accident; delivers a lecture on interior design for the Japanese jumbo jet at a meeting of the Japan Industrial Designers Association; serves as a permanent committee member of the Japan Industrial Design Promotion Organization.

1971	**End of the gold standard for American currency.**
	Backed by American air and artillery support, South Vietnamese troops invade Laos.
	China joins the United Nations.
	United States ends its trade embargo of China.
	The Japanese Red Army is founded by Fusako Shigenobu.
	Linda Nochlin publishes "Why Are There No Great Women Arists?"
	The Rothko Chapel in Houston, TX, designed by Mark Rothko and Philip Johnson, is completed.
	Louis I. Kahn wins the AIA Gold Medal.
	Noguchi begins designing the Horace E. Dodge Fountain and Philip A. Hart Plaza in Detroit, MI, with Shoji Sadao.
	KDA designs interiors and furniture for the Keio Plaza Hotel, Tokyo.
	In May, Kenmochi receives an award from the Japan Industrial Designers' Association.
	June 3, Kenmochi commits suicide at his Tokyo office.
	June 10, a funeral is organized by Kenmochi Design Associates; Tetsuo Matsumoto becomes KDA president.
	June 18, Kenmochi posthumously receives the Order of the Sacred Treasure from the Emperor of Japan.
1972	**U.S. President Richard Nixon visits China.**
	Israeli athletes are taken hostage and killed at the Munich Olympics.
	Nguyen Hue offensive begins in Vietnam.
	Japan normalizes diplomatic relations with the People's Republic of China.
	Okinawa Island is returned to Japan.
	Three Japanese Red Army members kill twenty-four and injure more than seventy-five at Tel Aviv's Lod Airport (now Ben Gurion Airport).
	Richard Serra begins constructing steel sculptures.
	Leo Steinberg coins the term "postmodernist" to describe Robert Rauschenberg's work.
	Winter Olympics open in Sapporo, Japan.
	Noguchi creates sculpture for the Bayerische Vereins Bank in Munich, and has exhibitions at the Cordier and Ekstrom Gallery in New York, Gimpel Fils in London, and Gimpel und Hanover in Zurich.
1973	**Oil embargo on the U.S. begins; OPEC doubles the price of oil.**
	Egypt and Syria attack Israel in Yom Kippur War, ending after twenty days.
	Massive protests against the Vietnam War continue in Washington, D.C.
	Sears Tower in Chicago by Skidmore, Owings, and Merrill is finished, making it the tallest building in the world.

Vietnam Peace Pact signed.

The Conference of International Council of Societies of Industrial Design (ICSID '73) is established.

Issey Miyake is represented at the Paris collections for the first time.

Artists Pablo Picasso and Robert Smithson die.

From January to February 4, 1973, Noguchi's Akari are included in the exhibition *Art is Meaning–Art in Use* **at Washington Art Gallery, Wykeham Rise School, CT.**

Noguchi exhibits sixteen recent sculptures at the Minami Gallery, Tokyo.

In November, Noguchi's Doncho (a theatre curtain) is included in *Exhibition of Kawashima Tapestries* **at Palace of the Peace, The Hague, Netherlands.**

The exhibition *Kenmochi Isamu: His Ideas and Creations* **opens at Matsuya Department Store, Tokyo.**

Five volumes of the book *The World of Kenmochi* **are published.**

1974 **U.S. President Nixon resigns and is pardoned by President Gerald R. Ford.**

The Japanese Red Army seizes the French Embassy in The Hague, Netherlands.

India tests its first nuclear bomb.

I. M. Pei begins designing the East Wing of the National Gallery in Washington, D.C. (completed in 1978).

Physicist Stephen Hawking proposes a new theory of black holes.

Eisaku Sato wins the Nobel Peace Price for guiding Japan into the Nuclear Non-Proliferation Treaty.

Noguchi collaborates with Shoji Sadao on Intetra Fountain in Palm Beach, FL; Noguchi designs six granite fountains for the Supreme Court Building in Tokyo; begins Shinto for Bank of Tokyo Building in New York; acquires building across the street from his Long Island City studio and starts renovation for display and storage of his sculpture. *Void* **is installed at Pepsico, Purchase, New York, and the exhibition** *Jean Arp and Isamu Noguchi* **opens at the Cordier and Ekstrom Gallery.**

1975 **Civil war begins in Lebanon.**

The first G6 Summit is held in France.

Altair, the first home computer, is introduced.

Designed by Roche-Dinkeloo, the U.N. Plaza in New York is completed.

New York City is bailed out of bankruptcy by the federal government.

The Vietnam War ends as Communist forces overtake Saigon, and South Vietnam unconditionally surrenders.

Eight-thousand life-size ceramic warrior statues are discovered in the tomb of Chinese emperor Qinshihuang in Xi'an, China.

U.S. Apollo and the U.S.S.R.'s Soyuz dock successfully in space.

Noguchi creates *Landscape of Time* **for the Jackson Federal Building in Seattle, WA; exhibits steel sculptures at the Pace Gallery, New York; begins Playscapes playground for Piedmont Park in Atlanta, GA, with Shoji Sadao.**

1976 **Apple Computer Company is founded.**

Matsushita Electric Industrial Co., Ltd., introduces the VHS home video cassette recorder.

Mao Zedong dies.

Jimmy Carter is elected U.S. President.

Robert Wilson and Philip Glass present *Einstein on the Beach* **at Festival d'Avignon in France.**

Designed by I. M. Pei, John Hancock Tower in Boston, MA, is completed.

Artists Hans Richter, Alexander Calder, and Alvar Alto die.

Noguchi completes *Portal***, a large outdoor sculpture for Cuyahoga Justice Center, Cleveland, OH, and begins Sky Gate for Honolulu, HI, as well as a fountain for the Art Institute of Chicago. Designs Martha Graham Dance Theater and Friendship Fountain with Shoji Sadao for the Missouri River between Iowa and Nebraska, both unrealized.**

The exhibition *Anonymous Cultural Treasures Collected by Isamu Kenmochi* **opens at Matsuya Department Store, Tokyo.**

1977 **Treaty to sign over the Panama Canal is passed by U.S. Congress.**

Artists Space in New York mounts the exhibition *Pictures***, introducing Barbara Kruger, Cindy Sherman, and Sherrie Levine.**

Designed by Renzo Piano and Gianfranco Franchini, the Centre Georges Pompidou opens in Paris.

The World Trade Center in New York is completed.

Noguchi begins creating Momo Taro for Storm King Art Center, Mountainville, NY. The exhibition *Isamu Noguchi: The Sculptor as Designer* **is organized by The Museum of Modern Art, New York.**

Commissioned by architect Kenzo Tange, Noguchi begins designing an interior stone garden for the Sogetsu Flower Arranging School, Tokyo.

1978 **Sunshine 60 Building is completed in Ikebukuro, Tokyo, at that time the tallest building in East Asia.**

New Tokyo International Airport opens in Narita.

Philip Johnson wins the AIA Gold Medal.

Treaty of Peace and Friendship is signed between Japan and China.

Bad Painting **exhibition opens at the New Museum of Contemporary Art, New York, featuring artists who reject Minimalism and Conceptualism.**

Noguchi's *Imaginary Landscapes* **exhibition, organized by the Walker Art Center, Minneapolis, MN, opens and travels to the Detroit Institute of the Arts, the Philadelphia Museum of Art, and the Whitney Museum of American Art through 1980.**

Noguchi exhibits Akari at the University of Hawaii Art Gallery, and designs the Lillie and Roy Cullen Sculpture Garden for The Museum of Fine Arts, Houston, TX (completed in 1986).

Sam Hunter publishes *Isamu Noguchi***, the first book-length monograph on Noguchi.**

1979 **Sony Corporation releases the first Walkman.**

I. M. Pei wins the AIA Gold Metal.

The Xanadu Houses, a series of experimental homes built to showcase computers and automation in the home, are designed in the U.S.

Ryohin Keikaku Co., Ltd., establishes Muji, a company that manufactures low-priced products with simple designs, natural materials, and plain packaging.

Noguchi begins plans with Shoji Sadao for Bayfront Park, Miami, FL.

August 4 to September 30, Akari Model # K4 is shown in *Recent Acquisitions: Architecture and Design* **at The Museum of Modern Art, New York.**

1980 **Ronald Reagan wins the U.S. presidency.**

Israel and Egypt establish diplomatic relations.

Iran-Iraq War begins.

Luis Barragan wins the Pritzker Prize.

Japan Style **exhibition opens at Victoria and Albert Museum, London.**

Akari are exhibited in *Isamu Noguchi and Hiroshi Teshigahara* **at the Sogetsu Kaikan, Tokyo.**

The Sculpture of Spaces, **an alternate version of** *Imaginary Landscapes*, **opens at the Whitney Museum of American Art.**

Noguchi initiates redesign of *Bolt of Lighting . . . Memorial to Ben Franklin*, **Philadelphia, PA (installed, 1984); begins** *California Scenario* **for South Coast Plaza, Costa Mesa, CA,** *Constellation (For Louis I. Kahn)* **at the Kimball Art Museum, Forth Worth, TX, and** *To the Issei* **and the Japanese-American Cultural and Community Center (JACCC) Plaza in Los Angeles, CA.**

Noguchi's *Shinto* **at the Bank of Tokyo in Lower Manhattan is destroyed by a new developer.**

Installation of *Passage of Seasons* **at the Cleveland Art Museum and** *Unidentified Object* **at The Metropolitan Museum of Art, New York.**

1981 **The Israeli Air Force destroys Iraq's Osirak nuclear reactor.**

Egyptian president Anwar Sadat is assassinated.

The first launch of Space Shuttle Columbia.

Marcel Breuer and Wallace K. Harrison die.

Richard Serra's *Tilted Arc* **is installed at the Federal Plaza in Lower Manhattan.**

The first IBM personal computer goes on the market.

Maya Ying Lin wins the public design competition for the Vietnam Veterans Memorial, Washington, D.C.

Arata Isozaki, apprentice to Kenzo Tange, begins a design for the Museum of Contemporary Art, Los Angeles (completed in 1986).

Noguchi purchases land next to his Long Island City building, and with Shoji Sadao begins design of The Isamu Noguchi Garden Museum; works on a series of galvanized-steel sculptures with Gemini G.E.L., Los Angeles. *Granits, basaltes, obsidiennes* **exhibit opens at Galerie Maeght, Paris.**

1982 **The Vietnam Veterans Memorial is erected in Washington D.C.**

Mexico's economy collapses.

Philippe Starck designs the interior for the private apartments of French President François Mitterand.

New technologies such as word processors, fax machines, microwaves, and satellite and cordless phones become common.

Noguchi exhibits Akari at Kasahara Gallery, Osaka. *California Scenario* **opens in Costa Mesa, CA.**

1983 **A special commission of the U.S. Congress releases a report critical of the practice of interning Japanese during World War II.**

Tokyo Disneyland opens.

Microsoft Word is released.

R. Buckminster Fuller dies of a heart attack in Los Angeles.

I. M. Pei wins the Pritzker Prize, and Norman Foster wins the Royal Gold Medal.

Noguchi begins construction of the garden at his Mure-cho studio in Japan.

To the Issei **and the JACCC Plaza open in downtown Los Angeles.**

1984 **Ronald Reagan is elected to another term as U.S. President.**

The People's Republic of China and the United Kingdom sign the Sino-British Joint Declaration on the future of Hong Kong.

The Space Shuttle Discovery lands after its maiden voyage.

Macintosh is released by Apple Computer.

Designed by Richard Rogers, Lloyd's Building in London is completed.

Richard Meier wins the Pritzker Prize.

A methyl isocyanide leak from a Union Carbide Plant kills thousands in Bhopal, India.

Noguchi completes water garden for Domon Ken Museum, Sakata, Japan; celebrates his eightieth birthday at the Sogetsu Flowering Arranging School, where Akari are exhibited.

1985 **The Spiral Building by Fumihiko Maki opens in Aoyama, Tokyo.**

Takao Doi, Mamoru Hohri, and Chiaki Mukai are chosen to become Japan's first astronauts.

Tsukuba World Expo opens in Ibaraki, Japan.

Charlotte Perriand (1903–1999) has a retrospective of her work at Musée des Arts-Décoratifs in Paris.

Noguchi officially opens The Isamu Noguchi Garden Museum in New York; exhibits *Space of Akari and Stone* **at Yurakucho Art Forum, Tokyo;** *Isamu Noguchi: Bronze Sculpture, 1959–1962* **opens at the Arnold Herstand Gallery, New York;** *Isamu Noguchi: Stones* **opens at Kasahara Gallery, Osaka.**

1986	**The U.S.S.R. launches the Mir Space Station.** **U.S. space shuttle** *Challenger* **explodes, killing seven astronauts.** **Artist Henry Moore dies.** **Nuclear reactor explosion at Chernobyl, U.S.S.R.** **Takako Doi, the first female party leader in Japanese history, becomes president of the Japanese Socialist Party.** **Noguchi exhibition** *Seven Stones* **opens at the Pace Gallery, New York; Noguchi represents the United States at the Venice Biennale with** *Isamu Noguchi: What Is Sculpture?* **Noguchi receives the Kyoto Prize from Inamori Foundation, Japan.**
1987	**Japanese National Railways is privatized and becomes Japan Railway Company.** **Construction begins on Kansai International Airport, built on an artificial island in Osaka Bay.** **Noguchi is distinguished with the National Medal of Arts, presented by President Ronald Reagan in Washington, D.C.**
1988	**U.S. President Ronald Reagan apologizes for the internment of Japanese Americans in camps during World War II.** **Seikan Tunnel opens, connecting the main island of Honshu with the northern island of Hokkaido.** **In April the Great Seto Bridge opens, connecting Honshu Island with the island of Shikoku, near Noguchi's studio in Mure-cho.** **Exhibition** *Isamu Noguchi: Bronze and Iron Sculpture* **opens at the Pace Gallery, New York.** **Noguchi is awarded Third Order of the Sacred Treasure by the Japanese Government, and is asked to design a master plan for Moere Numa Park in Hokkaido, Japan.** **Noguchi is recognized with an Award for Distinction in Sculpture from the Sculpture Center, New York.** **Noguchi dies from pneumonia on December 30 at New York University Hospital, New York.**

AUTHORS

JENNY DIXON was appointed Director of The Isamu Noguchi Foundation and Garden Museum in April 2003. Dixon began her arts career in 1977 when she joined New York City's Public Art Fund, subsequently serving as Executive Director from 1980 to 1986. At the Fund she assisted numerous artists in the realization of their public works. It is in this capacity that she came to know and work with Isamu Noguchi. Among her many accomplishments at the Fund was the initiation of the New York City "Percent for Art" program. Dixon's previous positions include Executive Director of the Lower Manhattan Cultural Council and Director of The Bronx Museum of the Arts. She has taught at Cooper Union, Parsons School of Design, and New York University's Steinhardt School.

TETSUO MATSUMOTO is an architect and president of Kenmochi Design Associates (KDA). Upon graduation from the Department of Architecture at Chiba University, Matsumoto entered the Industrial Arts Research Institute as an engineer. During his tenure at KDA, which began in 1957 as Chief Designer, Matsumoto worked on the interior designs of the Japanese Pavilion at the World Expo in Brussels (1958) and the Kyoto International Conference Center (1966). His recent work includes a restaurant at Hotel Nikko Saipan (1989), and the Shinkansen train coach E2 (2000) and N700 (2007).

NINA MURAYAMA is a Ph.D. candidate in Art History at The Graduate Center, The City University of New York. Murayama's studies focus on twentieth-century art and design. She received a B.A. in Engineering at Chiba University, and is currently working on a doctoral dissertation on Donald Judd's furniture.

KATARINA V. POSCH is an associate professor of design history and theory at Pratt Institute, New York. Posch has curated exhibitions at the Pompidou Center in Paris and the Vitra Design Museum in Weil am Rhein. At the Vitra Museum she curated *Isamu Noguchi: Sculptural Design*, an exhibition designed by Robert Wilson that traveled to The Noguchi Museum. Awarded a Ph.D. from the National University of Fine Arts and Music in Tokyo, Japan, she has published and lectured widely on cross-cultural and cross-disciplinary topics in the field of design history.

BONNIE RYCHLAK is the curator at The Noguchi Museum, having previously served as registrar and director of collections. An artist herself, Rychlak worked as Noguchi's assistant from 1980 until the artist's death in 1988, and is an authority on Noguchi's vast body of work. She has curated and written catalogue essays for numerous exhibitions at The Noguchi Museum and other institutions, including *Zen No Zen: Aspects of Noguchi's Sculptural Vision*, *Noguchi: The Bollingen Journey, Photographs and Drawings, 1949–1956*, and *Noguchi and the Figure*. Her accomplishments include a Rockefeller Foundation residency at the Bellagio Study and Conference Center and the Rome Prize from the American Academy in Rome.

YOSHIRO TANIGUCHI (1904–1979) was a pioneering architect and scholar, whose work and influence transformed the environs of Tokyo. Born in Kanazawa, Japan, he graduated from the Architecture Department of Imperial University. In addition to *Shin Banraisha* at Keio University, his architectural achievements from the 1960s onward include the Imperial Theater, the Tokyo National Museum of Modern Art, and the Peace Pagoda in San Francisco.

MOSUKE YOSHITAKE (1909–1993) graduated in Metal Arts from the Tokyo Fine Arts School in 1935. After his studies, he worked at the Industrial Arts Research Institute and later became a Professor at the Musashino Art University and director of the Japan Designer-Craftsman Association.

PHOTO CREDITS

Every effort has been made to trace and correctly credit the source and ownership of all copyrighted material included in this volume. Any errors or omissions are inadvertent and will be corrected in subsequent editions, provided notification is given to the publishers.

Unless otherwise noted below, all images are courtesy of The Isamu Noguchi Foundation and Garden Museum.

Front cover and page ii, The Museum of Modern Art, New York, The Philip L. Goodwin Collection (gift of James L. Goodwin, Henry Sage Goodwin, and Richmond L. Brown) (107.1958)/Licensed by SCALA, Art Resource, NY; **page iii**, courtesy of Tendo Co., Ltd.; **front cover and page v**, courtesy of Kokushokankokai, Inc.; **front cover and page vi**, photograph by R. Martin; **page vii**, photograph by Yasuhiro Ishimoto, courtesy of Kenmochi Design Associates (KDA); **page 12**, bottom, photograph by Sasha Stone; **page 13**, photographs by Kevin Noble; **pages 14–17**, photographs by Ezra Stoller © Esto; **page 20**, photograph by Kevin Noble; **pages 22–23**, courtesy of the National Institute of Advanced Industrial Science and Technology, Japan (AIST), http://aist.go.jp; **page 24**, top, photograph by Ezra Stoller © Esto; **page 24**, bottom, photograph by Kevin Noble; **page 25**, bottom, photograph by Kevin Noble; **page 26**, top, photograph by Tom Leonard; **page 27**, photographs by Kevin Noble; **page 30**, top, courtesy of Industrial Arts Foundation; **page 30**, bottom, courtesy of the Matsudo City Board of Education, Chiba Prefecture, Japan; **page 31**, top, courtesy of Industrial Arts Foundation; **page 32**, courtesy of Avery Architectural and Fine Arts Library, Columbia University, © Artists Rights Society (ARS), New York/ADAGP, Paris; **page 33**, courtesy of The Frank Lloyd Wright Foundation, Taliesin West, Scottsdale, AZ; **page 35**, top left, photograph by Yasuhiro Ishimoto, courtesy of M.I.T. Press; **page 35**, top right, photograph by Shinchosha Publishing Co., courtesy of M.I.T. Press; **page 35**, bottom left and right, and **page 36**, courtesy of Industrial Arts Foundation; **page 37**, top, courtesy of Taro Okamoto Museum, Kawasaki; **page 37**, middle, courtesy of Konan School Public Relations Department, Hyogo, Japan; **page 37**, bottom, courtesy of Taro Okamoto Museum, Kawasaki; **pages 38–39**, photographs by Nomadic Studio; **page 43**, courtesy of Time Life Pictures/Eliot Elisofon/Getty Images; **page 44**, top, courtesy of Tokyo National Museum/TNM Image Archives; **page 45**, top, courtesy of Tokyo National Museum/TNM Image Archives; **page 46**, photograph by Yasuhiro Ishimoto; **page 47**, courtesy of Tendo Co., Ltd.; **page 48**, bottom, courtesy of Shigeru Yura; **page 49**, courtesy of Kotobuki Corporation; **page 50**, courtesy of Industrial Arts Foundation; **page 51**, courtesy of the Matsudo City Board of Education, Chiba Prefecture, Japan; **page 53 and back cover**, photograph by Michio Noguchi; **page 54**, top, and **page 55**, photographs by Erik Johnson; **page 55**, bottom, photograph by Kevin Noble; **page 56**, courtesy of Kokushokankokai, Inc.; **pages 57–59**, photographs by Kevin Noble; **front cover and page 62**, courtesy of Industrial Arts Foundation; **pages 64–65**, courtesy of Industrial Arts Foundation; **pages 66–69**, courtesy of the Matsudo City Board of Education, Chiba Prefecture, Japan; **page 70**, photograph by Nomadic Studio; **page 71**, top, photograph by Shufuku Kenkyujo; **page 71**, bottom, photograph by Nomadic Studio; **page 72**, courtesy of Industrial Arts Foundation; **page 73**, top, photograph by Michio Noguchi; **page 73**, bottom, photograph by Chuji Hirayama; **page 74**, top and bottom, and **page 75**, courtesy of Industrial Arts Foundation; **page 76**, photograph by Eliot Elisofon, courtesy of Time Life Pictures/Eliot Elisofon/Getty Images; **page 77**, photographs by Kevin Noble; **pages 79–81**, photographs by Kevin Noble; **pages 86–89**, photographs by Kevin Noble; **page 90**, top, photograph by Hans Namuth, courtesy of the Center for Creative Photography, University of Arizona © 1991 Hans Namuth Estate; **page 90**, bottom, photograph by Dan Budnik; **page 91**, top, photograph by Atelier Stone; **page 91**, bottom, cover and interior pages reprinted by permission from *What is Modern Design?* © 1950 The Museum of Modern Art, New York, and interior reproduction, page 26, courtesy of Edvin Ohrström © 2007 Artists Rights Society (ARS), New York/BUS, Stockholm; **page 92**, reprinted by permission from Bloomingdale's, New York; **page 93**, bottom left, photograph by Tatsuo Kondo; **page 93**, bottom right, photograph by Taylor and Dull; **page 94**, top, photograph by Andrew Popper; **page 95**, photograph by Shigeo Anzai; **page 96**, bottom, photograph by Eliot Elisofon, courtesy of Time Life Pictures/Eliot Elisofon/Getty Images; **page 104**, courtesy of the Matsudo City Board of Education, Chiba Prefecture, Japan; **pages 105–106**, photographs by Nomadic Studio; **page 107**, courtesy of the Matsudo City Board of Education, Chiba Prefecture, Japan; **page 108**, top, courtesy of Tendo Co., Ltd.; **page 108**, bottom, photograph by Nomadic Studio; **page 109**, courtesy of KDA, and published in Tetsuo Matsumoto et al., *The World of Kenmochi Vol. 4, Chronology* (Tokyo: Kawade Shobo Shinsha Publishers, 1975), page 166; **page 112**, top, courtesy of the Matsudo City Board of Education, Chiba Prefecture, Japan; **page 113**, photograph by and courtesy of Eiji Fujimoto; **page 114**, courtesy of Industrial Arts Foundation; **page 115**, photograph by Kazuo Usui; **page 116**, top, courtesy of Y.M.K. Co., Ltd.; **page 116**, bottom, courtesy of the Matsudo City Board of Education, Chiba Prefecture, Japan; **page 117**, courtesy of KDA, and published in Tetsuo Matsumoto et al., *The World of Kenmochi Vol. 4, Chronology* (Tokyo: Kawade Shobo Shinsha Publishers, 1975), page 168; **page 118**, top, photograph by Osamu Murai; **page 118**, bottom, courtesy of Kokushokankokai, Inc.; **page 119**, photographs by Nomadic Studio; **page 120**, courtesy of Tendo Co., Ltd.; **page 121**, top, courtesy of Tendo Co., Ltd.; **page 121**, bottom, photograph by Nomadic Studio; **page 122**, top, photograph by Yasuhiro Ishimoto, courtesy of the Matsudo City Board of Education, Chiba Prefecture, Japan; **page 123 and front cover**, photograph by Chu Fujiwara, courtesy of Kyoto International Conference Center; **page 124**, courtesy of Ohtone Country Club; **page 125**, photographs by Nomadic Studio; **pages 126 and 128–136**, reproductions of *Industrial Art News*, vol. 18, no.10, October 1950, courtesy of AIST; **pages 138 and 140–141**, reproductions of *Industrial Art News*, vol. 22, no. 9, September 1954, courtesy of AIST.

INDEX

Five Ties Publishing, Inc., New York
www.fiveties.com

Produced and edited by Garrett White
Associate Editors: Genevieve Cortinovis, Amanda Thorpe
Book design: Garrett White

First edition 2007

© 2007 The Isamu Noguchi Foundation and Garden Museum / Five Ties Publishing, Inc.
www.noguchi.org

Printed by Graphic Management, 270 Sheffield Street, Mountainside, New Jersey 07092

All rights reserved. No part of this book may be reproduced in any form by any electronic or mechanical means (including photocopying, recording, or information storage and retrieval) without permission in writing from the publisher.

Printed in the U.S.A.

ISBN 978-0-9794727-0-1